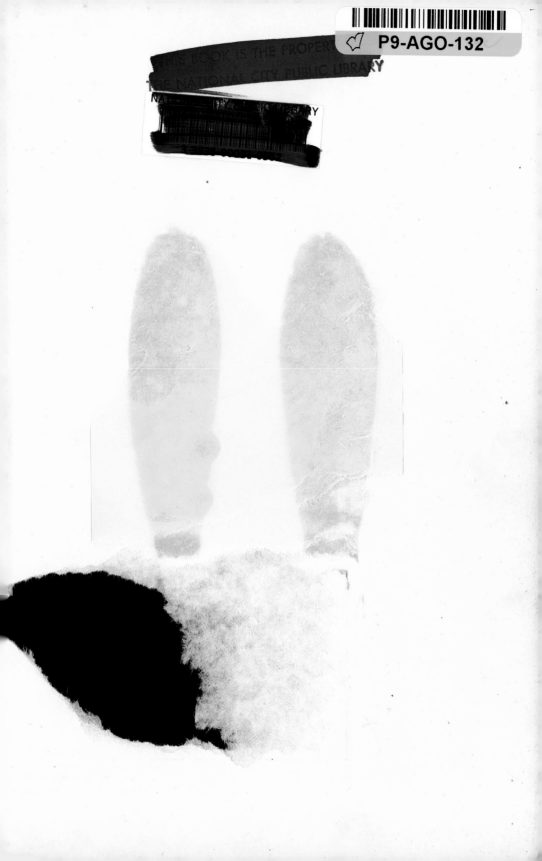

A Capitalist Romance

A Capitalist Romance:
SINGER AND THE SEWING MACHINE

by
Ruth Brandon

J. B. LIPPINCOTT COMPANY
PHILADELPHIA AND NEW YORK

For
PHIL

Copyright © 1977 by Ruth Brandon
All rights reserved
First edition
Printed in the United States of America
9 8 7 6 5 4 3 2 1

U.S. Library of Congress Cataloging in Publication Data

Brandon, Ruth.
  A capitalist romance: singer and the sewing machine.

  Bibliography: p.
  Includes index.
  1. Singer, Isaac.   2. Sewing-machine industry—United States—
History.   3. Inventors-United States—Biography.   I. Title.
TJ140.S59B7   338.7′68′17677   76-55725
ISBN-0-397-01196-2

Contents

List of Illustrations

A sixteen-page section follows page 114.

Preface

For most biographers, letters from, to and about their subject, to say nothing of journals if the subject happened to be a journal-keeping sort of person, form an important part of the source material. But no materials of this kind are available to the biographer of I. M. Singer. Singer was not a letter-writing man, since he had some difficulty in writing anything at all, and the idea of his keeping a journal is simply laughable. It is therefore very difficult to get at his life, so to speak, from the inside.

The biographer, in this situation, is in a position similar to that of the physicist who must draw evidence for the existence of certain particles from the traces of their impact upon other particles. Singer was a forceful and flamboyant man and his impact on the world around him was considerable. Moreover the last part of his life, both personal and professional, was passed in a blaze of publicity, matched only by the obscurity in which he passed his first forty years. The biographer's job is therefore a delicate one. Evidence relating to his early life is scant, and that on his later life is almost all slanted. People did not generally record their opinions of and encounters with I. M. Singer in a void; they did so usually to achieve some specific end: to gain or regain money, to win a lawsuit, to obtain a divorce. The only persons who might have provided some disinterested reminiscences were his children, but most of them had—as will emerge—their own reasons for not drawing too much attention to their spectacular parent and all are now long since dead. There is only one surviving grandchild that I know of. There are large numbers of great- and great-great-grandchildren, but none of these was acquainted with I. M. Singer himself. It was therefore necessary to try and reconstitute Singer from press clippings, company records, memoirs by acquaintances, legal testimony and a few family letters and records. The task has been a fascinating one because Singer's life was so extraordinarily interesting in so many respects (both historical and scandalous) and covers such an enormous spectrum of society in America and Europe during that heyday of Progress, the mid-nineteenth century. Singer himself embodied many of the myths of that era: the poor immigrant's son who makes good in the new world, the self-made man who tries to storm the bastions of society. His story is also of considerable interest to commercial and technical history. For these reasons it is curious that nobody has published a biography of him in the hundred years which have elapsed since his death. It seems clear that this has been at least partly because neither the family nor the company wanted to advertise old and (they hoped) dead scandals. But times change, and after so many years such considerations are scarcely relevant. I have in fact received the greatest possible help and cooperation from both the Singer Company and the Singer family, for which I am extremely grateful.

I should particularly like to thank the following individuals who have all helped most generously during the preparation and writing of this book: Anson Burlingame; A. R. Boyle and Arthur Dorman of Singer's at Clydebank; Caroline and Ian Davidson; Elie Duc Decazes; C. M. Eastley; Jacques Ehrsam; Crispin Gray; Brian Jewell; Dick Kosmicki and Bogart Thompson of Singer's in New York; Winnaretta Lady Leeds; Jan Marsh; John and Erni Meyer; Michael Motley; Alain Ollivier; Forbes Singer; Diana Vincent Daviss; and Philip Steadman. I should also like to thank members of the staff of the New York Public Library; the New-York Historical Society; the Science Museum, South Kensington; the Science Museum Library; Torquay Public Library; and the Wisconsin State Historical Society Library, Madison.

For permission to quote from published material I would like to thank the Hutchinson Publishing Group Ltd. (UK) and The Dial Press

(USA) (*The Real Isadora*, Victor Seroff, 1972), Hart Davis MacGibbon (UK) and Farrar, Straus & Giroux, Inc. (USA) (*The Incredible Mizners*, Alva Johnston, 1953), and Chatto and Windus Ltd. (UK) and Little, Brown and Company in association with the Atlantic Monthly Press (USA) (*Marcel Proust*, vol. 2, George Painter, 1965).

Prologue

The 1850s saw the greatest mass of immigrants ever known flood into America from the poorer parts of Europe—from Germany, from Scandinavia, and most of all from Ireland. Tens of thousands poured in every year, leaving behind famine, exploitation and the most primitive of living conditions, and expecting—well, what were they expecting? After the rigors of the dreadful crossing, cooped up in steerage with as many others as the captain could possibly squeeze on to his ship inside or outside the regulations, it is perhaps doubtful whether on landing many of them could have formulated anything so coherent as an expectation. Did they believe that wide-open spaces were there for the taking? There were indeed such spaces to be had; but they were far from New York and Boston, where the immigrants landed; and clearing virgin lands required a toughness and resource of body and spirit which the undernourished and demoralized immigrants simply did not possess. Besides, the assembled sharks who met the immigrant ships were there largely to ensure that the new arrivals did not leave the city for the continent beyond. Touts for boarding-houses would gather up a family's bags and belongings—and disappear with them. Self-appointed guides from the immigrants' homeland would collect their money for the journey by railway and canal boat to some far-off destination—and leave the party stranded, if not in the port itself, then not far outside it.

Thousands and thousands of these poor immigrants were thus left without money, work or prospects in the cities of the eastern seaboard—above all, in New York. If they had believed the tales back home about the streets of New York being paved with gold, the realities of life in the city disabused them of such notions within a very few days. There were slums in New York to rival any city in the world. The Five Points, between Broadway and the Bowery, might, said Dickens in his *American Notes*, "in respect of filth and wretchedness . . . be safely backed against Seven Dials, or any other part of famed St. Giles's." Moreover, there was not much to distract them from their living conditions; for utilitarian New York did not boast the street amusements—bands, conjurers, barrel-organs and the rest —which kept the meaner streets of Europe so lively. But this is not to say that there was nothing to look at in New York, nor anything to amuse the idle spectator. For anybody, of course, was free to walk up to fashionable Fifth Avenue, and there, indeed, were sights to be seen.

Imagine, then, the immigrant family, just arrived from the old world, taking a walk—why not?—around their new hometown. They are probably dazed and despondent and quite unclear as to why they have subjected themselves to this dismal and shattering upheaval.

Along the avenue comes the answer to this very question: a vast and shining equipage drawn by nine matched horses. It is painted a bright canary yellow, trimmed with black. On the outside are seats for a small orchestra, which is playing merrily despite the jolting. Inside ride a number of fashionably dressed ladies and gentlemen. From the back emanate the shouts and squalls of the very youngest members of the party, for the coach is conveniently fitted with a nursery compartment. Everyone knows the proprietor of this impressive contraption. It is Isaac Merritt Singer, the sewing machine millionaire.

To the gasping spectators, such show represents wealth beyond their wildest dreams. Yet that, of course, is why they are here: in this country, people like them can become people like that. Or so it is said. Singer might be the very embodiment of their dreams. Ten years ago, he was as poor as they are now. Moreover, he is the son of an immigrant himself. His own father, just arrived from Germany, must have felt very much as they do now.

It is men like Singer who represent the reason why they are now standing bewildered in this strange street in an alien land. If they do not make it themselves, there is a good chance that their sons will do so, as Adam Singer's son so indisputably has.

I

Early Influences

Little is known of Adam Singer, the father of Isaac Merritt Singer. This is not surprising: he was an obscure immigrant from an obscure family, doing his best to make a new life in a new country. Such people rarely leave copious personal records; they have usually neither the time, ability nor inclination to do so.

Such information as we have about Adam Singer comes from three sources, the most reliable of which is probably William Singer, Isaac's oldest son. He retailed what he knew in a series of letters to his half-sister Winnaretta, by then Princesse de Polignac, written in 1905 and 1906 long after both their father and grandfather were dead. Being thirty-one years older than Winnaretta, William had had considerably more acquaintance with their father, and he had also known their grandfather, which she had not. About Isaac he was rather reticent. "It would deeply interest you to know *all* his wonderfull and very *unusual* history, but *it better not be on paper*. When we meet I will tell you if you wish," he wrote. About Adam, however, he was more forthcoming. "Our grandfather Adam Singer was born in the year of 1753, and when he was sixteen years old in 1769, he came *alone* to New York, U.S.A. and never thereafter left America; he finally settled in Troy, State of New York, and there in the year 1788 he married his *first* wife, an *American* woman whose foreparents nearly a hundred years before came from Holland." In another letter, William enlarged on this:

Grandfather Adam was the youngest one of a *very large family of brothers in Saxony.* Adam realizing that chances were few for *him*

5

there, he all alone came to New York. The exact year no one *really* knows, our father himself did not *know* for *certain*, but if father was *right* that grandfather lived *102* years, then by estimation he arrived here in 1769, he being then 16 years of age. Arriving *stranger and no knowledge* english—he came alone. In course of time he drifted to Troy this State, where he went into the cooperage business, a great business at that time, far *more* so than *now* . . . In Troy he married a Troy lady of Holland extraction and descent, he being then 35 years of age, and 42 years after, after a large family were born to him, *our father was born Oct. 26th 1811*. [The forty-two years refers to the time elapsed from Adam's landing on the shores of the New World.]

There are various theories as to Adam's original extraction. Many people assumed, since Singer is frequently a Jewish name, that the family must originally have been Jewish. One of his grand-daughters★ found some evidence to support this theory after Singer's death. She was spending a year at school in Germany, and there she met a girl from Frankfurt who told her that this had been the original hometown of the Singers. The facts were naturally more talked about, she said, now that the Singers had become so famous. In fact it was believed that the family name was originally Reisinger, a common Jewish name, and the Reisinger family was indeed Jewish. Adam's father, however, had married out of the faith, and Adam was brought up a Protestant like his mother. It seems possible then that Adam did, as William says in one of his letters, come to America "because of his protestant convictions"—there may have been pressures on him from his father's side of the family. At any rate, his father agreed to the plan and gave him enough money to pay the fare and provide him with a start in his new country. After his arrival (this version continues) Adam shortened his name from "Reisinger" to "Singer" as he thought this would sound less Germanic and be easier to pronounce.

There is another theory that Adam was a Mennonite, and joined the Mennonite emigration from Germany to America. This, however, would seem to be rather unlikely, given William's insistence that his

★ For source references and additional information, see Notes, pp. 223–32.

grandfather came alone. The rather unsettled nature of his family life also seems to weigh against this theory, since a close-knit family group has always been a characteristic of life in Mennonite and similar communities.

Whatever the truth, it is hard to imagine that such details ever held any importance for Adam's son Isaac. Various religions were at different times ascribed to him. When he died, special mention was made of the fact that he received communion on his deathbed, thus confounding those who had tried to brand him an atheist. The *National Cyclopedia of American Biography* says that he was Episcopalian, but this tells us little more than that various members of his family nurtured ardent desires to seem respectable. But no such desire had ever manifested itself in Isaac and it is unlikely that he ever gave a voluntary thought either to respectability or to religion. Indeed, his career is notable, among other things, for these two salient omissions, neither of which was likely to go unremarked in respectable mid-Victorian circles.

Adam Singer's trade, although William said he was a cooper, is in fact generally given as millwrighting. Not long after the birth of their youngest child, Isaac Merritt Singer, in the village of Schaghticoke near Troy, Adam, his wife Ruth, and their family moved 150 miles west, settling in Oswego on the shores of Lake Ontario. It was here that Isaac spent his childhood.

Oswego lay in the midst of that part of New York State which had been the territory of the Six Nations of the Iroquois until 1788 when a settlement as to the purchase of lands was reached between the old inhabitants and the new. In September of that year, the Onondaga tribe (Oswego is situated in Onondaga County) sold their land to the State of New York for "$1,000 in hand, and an annuity of $500 for ever." Other tribes were bought out on similar terms.

Only after this apparently equitable arrangement could settlers feel safe to move into the territory. One of the first of these, Asa Danforth, with all his family, accompanied by Comfort Tyler, "progressed far beyond the bounds of civilization, locating at Onondaga Hollow." There were no roads in this part of the country, so the party came by water, landing at the mouth of Onondaga Creek. Major

Danforth's wife saw no white woman in the first eight months they were settled there.

Twenty-five years later, when Adam Singer and his family arrived in Oswego, things had progressed considerably. There was now a recognizable town, or at least a village, with its church, its main square, and even its own newspaper, the *Oswego Palladium*, although civilization had not yet gone so far that subscribers to this last could not pay in kind—for example cords of wood, which they were urged to deliver to the newspaper offices without delay.

If Adam Singer was a millwright by trade, then from the business point of view the move was a sensible one. At this time, before the settlement of the great wheatlands of the Midwest and the opening of communications across the continent, this was an important wheat-growing area. A traveler in this area in 1830 remarked that "the weight of flour made here annually is prodigious." In other occupations, activity and wages were booming. "Mechanics are well paid for their labour," wrote a traveler visiting Batavia, New York, in 1811, the year Isaac was born.

Carpenters have one dollar a day, and their board; if they board themselves, one dollar and a quarter. Other trades have in proportion, and living is cheap. Flour is about five dollars per barrel; beef four cents per pound; fowls twelve and a half cents each; fish are plenty and cheap. A mechanic can thus earn as much in two days as will maintain a family for a week, and by investing the surplus in houses and lots, in a judicious manner, he may accumulate money as fast as the farmer, and both may be independent and happy.

Up to a point, the Singer family followed this advice. They had a house and a plot of land in the township of Granby, near a small creek on the Oswego River. In 1818 John V. Singer, one of Isaac's older brothers, paid fifty dollars for a parcel of land adjacent to his father's house; however, this was perhaps not quite the type of judicious investment recommended, since in 1819 he resold it for only forty dollars. It seems as if the family fortunes were having their ups and downs.

Indeed, despite the blithe assurances of the many emigrants' guidebooks of the period, life in America at this time was very far from being an assured success for all who undertook it. What the guides did not point out was that although prosperity was indeed there for the taking, not everybody was capable of taking it in the particular form in which it presented itself. Virgin land was cheap— it cost at this time two or three dollars an acre—but pioneering, the art of clearing the wilderness in order to make new farms, was a highly specialized accomplishment requiring immense perseverance, adaptability and practical knowledge. Where, as in Oswego and its surroundings, the first steps had been taken already, land was naturally much more expensive to buy.

In business, too, adaptability seems to have been, if not an absolute necessity as in pioneering, then a highly desirable quality. It becomes quite obvious, if one looks through the pages of the weekly *Palladium*, who the business leaders of the community were. The same names crop up again and again, singly or in partnership, forever trying something new: starting a grocery business when special lines of sugar or coffee were to be had, advertising them, winding up the business when they ran out, starting up again in partnership as forwarding agents in the developing port, advertising singly, as land agents—these were the people who would make the most of the business opportunities offered by the frontier. Again, if we are to judge by the *Palladium*, plenty of people did spectacularly badly; every week it carried at least one notice of a sheriff's sale of the seized lands and effects of some local bankrupt.

Adam Singer's name appears neither as a bankrupt nor as a business booster—indeed, the only trace to be found of the Singer family in the *Palladium* is when they failed to collect letters addressed to them which were waiting at the post office. However, the Singer household in Oswego was clearly not a happy one, for in 1821, when Isaac was only ten years old, Ruth divorced her husband and left home.

The usual image of family life in early America is of a very settled sort of affair—the hardworking wife standing staunchly by her pioneer husband. The pages of the *Palladium* and other such local newspapers, however, present us with quite a different picture. Here we frequently find notices put in by deserted wives in search of husbands who had disappeared, or husbands declaring that they will no longer be responsible for their wives' debts—"she having broken the marriage vows," as one put it. Divorce, however, was still a rare and scandalous occurrence, and Ruth, who came of pious Quaker stock, must have been hard pressed indeed to take such a drastic step. There was only one ground upon which a divorce could be granted in New York State at that time, and that was adultery. Sure enough, within a very short time Adam Singer married again. He was never again to see his first wife, who seems to have cut off contact with her whole family, retiring to a Quaker settlement in Albany. In 1852, Adam, then ninety-nine years of age, tried to find her once more; he traced her to Albany, only to discover that she had died there, aged ninety-six, the year before. Perhaps he wanted to tell her about their youngest son's late great success in life. Adam himself died in the Western Reserve, Ohio, in 1855 at the age of 102.

In 1821, however, at the age of sixty-eight, Adam was still in his prime. Most of his children had by now grown up and left home, but Isaac was still living under the paternal roof—a situation he found increasingly difficult. Under the circumstances it would be hardly surprising if his loyalties lay with his mother; at any rate, he did not get on with his stepmother. When he was twelve, therefore, Isaac decided to take his life into his own hands. With nothing to keep him at home, he left—"without money, without friends, without education, and possessed of nothing but a strong constitution and a prolific brain"—and made his way from Oswego, which even reading between the biased lines of the *Palladium* gives a distinct impression of smallness and dullness, westward to Rochester, which although it had only been founded about twelve years previously was already something quite different. Rochester was a boom town, the boom on which it rode being that of the phenomenally successful Erie Canal which ran through its center. "It is, indeed, scarcely credible that in the period of eighteen short years a place of the present extent and importance of Rochester should have arisen from the wilds of a forest," wrote John Fowler in 1830.

. . . There are not only spacious and well-arranged streets, with corresponding stores and warehouses, and private residences of elegance and respectability; but, besides a court-house, gaol, and eleven churches, two markets, two banks, and several very excellent hotels, there is a museum, institute, an athenaeum, an arcade, a Vauxhall, public baths, reading-rooms, etc., etc., and a population of more than 13,000 souls! and, in the face of all this, there are even now the stumps of trees standing in some of the streets.

One advantage which Rochester possessed and which had been unavailable in Oswego was a number of schools. Recalling his early life for a newspaper interview in 1853 the adult Singer remembered that "schools at that day and in that region were seldom to be found, and consequently the incipient inventor was wholly without the advantages of education, so long as he remained under the paternal roof." In fact such a statement probably reflects more on the situation in the Singer home than anything else; we may well suppose that the parents were too occupied with their own affairs to spend much time with their youngest son, who consequently was left very much to his own devices. This lack of concern was an exception to the rule; even if there were no formal schools available, most parents managed to make sure that their offspring could at least read, since literacy was always very much at a premium in America. Among the first institutions to be set up in any good-sized village were a subscription library and a newspaper, and even if the parents were illiterate someone could always be found to teach the children. (Thus Peleg Redfield in his recollections of life near Utica, New York recalled: "In 1800, a log house had been vacated; we fitted it up and hired Elam Crane to teach a school. It was a mile from my house, and my boys used to go through the woods by marked trees.")

Isaac stayed in Rochester for the next seven years, probably living during this time with one of his older brothers. There is a family tradition that he was brought up by an older brother, and this would make sense: Rochester is not far from Oswego, so that it seems likely that more than one of the Singers might have moved there, where opportunities were so much greater. This would also have given

Isaac a definite reason for making his way there when he left home.

Once Isaac was established in Rochester, he succeeded in earning enough by laboring during the summer to enable him to spend the winter months in "a common school, where he obtained the rudiments of learning." Certainly he never learned to spell, nor to write with any great ease, as the few extant specimens of his handwriting amply show. However, he certainly learned to read and to enjoy reading. "The pressing necessity of laboring three-fourths of the year to secure a livelihood, of course prevented anything like systematic study; still, whenever any book, treating of mechanics or the arts, came in his way, he read it with avidity and attention," he told the newspaper reporter in 1853.

During this time it seems fair to assume that he led a haphazard sort of life, picking up jobs and education where he could. Then in 1830, with his nineteenth birthday approaching, he decided that he needed some sort of trade with which he could be sure of earning a living. Accordingly he entered a machinist's shop (where machines were made and repaired for craftsmen, farmers, etc.) as an apprentice. He did not, however, stay there long. After little more than four months he picked up his tools and moved back east to Auburn.

The reason Singer gave in his 1853 interview for cutting short his apprenticeship so early was sheer ability: "At the end of four months, he was so far a skillful artisan, that few would have supposed he had not served a full apprenticeship in the trade." We shall see that there was reason to believe this may not have been the whole truth. He was certainly an exceptionally gifted mechanic, as his subsequent career was to prove. On the other hand, nobody can learn a trade in four months, however gifted, and a more likely explanation is that he got bored and felt like moving on. He was, he said, "inclined to see as much of the world as was practicable," and as a start picked up a contract in Auburn to construct some lathe-making machinery. This, by luck or judgment, he executed satisfactorily; he was then in a position to move on, plying and learning his trade as he went.

It sounds a sketchy sort of plan, and it was an unsettled beginning to what was to be a most unsettled life. Singer's longest stay in any one place was the thirteen years he spent in New York City between 1850 and 1863—during which period he moved house four times and was continuously traveling both in America and abroad. Not only could he not feel happy tied to one place: the prospect of a continuing relationship with any one person seems equally to have irked him. He was conspicuously unable to form satisfactory and durable relationships, in bed, business or friendship, with members of either sex. The boring virtues of perseverance never interested him, while novelty was something he could never resist.

It is tempting to attribute this inability to settle down to the psychological rigors of his early life: the breakup of his parents' marriage, his mother's departure when he was still very young, the unsatisfactory home provided by his father and stepmother. Doubtless this had its effect on his subsequently spectacular marital and extra-marital history: it is a social workers' commonplace that one broken marriage begets others. But if we are children of our parents, we are also children of our times, and Isaac Singer was in many ways a classic product of the age and place in which he lived. If he was a restless man, early nineteenth-century America was a restless place. If he hankered after novelty, the times encouraged him to do so: in the age of progress, "newer" was a synonym for "better." If he was unable to settle down satisfactorily to anything, such an ability was not the *sine qua non* of a successful life it has since become—and especially not in America. For in the New World, movement was in the air. In Europe, there was a strong probability that you would live out the greater part of your life near where you were born. In America, this was not the case. It was more than likely that sheer chance had landed your parents wherever it was they happened to be living; why should you, too, not take your chance? Europe was full up, but America was empty except for some Indians—being rapidly dealt with—and it was waiting to be opened out. The concept of "home" was never so strong in the United States as in Europe—and in fact it still is not, or at least it is different. In America the word "home" is a synonym for "house"; it is a traveling concept, one which you carry around with you—your home is wherever you happen to be living. One might speak of a "development of new homes" in America; in England, such a phrase would be nonsensical, because a house, in England, is merely a "house"; "home" is an altogether broader concept, implying rootedness and long residence.

In Europe, to throw up your apprenticeship and learn your trade as you went would be regarded disapprovingly as less than craftsman-like, which in a society amply supplied with craftsmen would be a strong condemnation. Besides, who voluntarily took up an itinerant life in Europe? Gypsies, tinkers and peddlers, the customary travelers of Europe, were the outcasts of society. A solid citizen settled down and plied his trade. But in America things were quite different. Able craftsmen, with or without a journeyman's certificate, were much needed and in short supply. A willingness to strike out was regarded as proof of adventurousness and self-reliance. And what was the great advantage of staying in one place in a country where only six percent of the population lived in cities, with the rest mostly scattered in lonely farmsteads several days' or even weeks' journey from the nearest settlement of any size? The isolation of most people's lives at that time is perhaps hard to imagine today, when our problem is more often the opposite one of seeing only too much of our neighbors. But the homesteader's wife who described her enforced self-sufficiency to Frances Trollope in 1828—just about the time when Isaac was setting out on his travels—was the rule rather than the exception:

> The woman told me that they spun and wove all the cotton and woollen garments of the family, and knit all the stockings; her husband, though not a shoemaker by trade, made all the shoes. She manufactured all the soap and candles they used, and prepared her sugar from the sugar-trees [maples] on their farm. All she wanted with money, she said, was to buy coffee and tea, and whiskey, and she could "get enough any day by sending a batch of butter and chicken to market." They used no wheat, nor sold any of their corn, which, though it appeared a large quantity, was not more than they required to make their bread and cakes of various kinds, and to feed all their live-stock during the winter . . . "'Tis strange to us to see company; I expect the sun may rise and set a hundred times before I shall see another *human* that does not belong to the family."

In such a society, the itinerant fulfilled a real need—or, rather, he fulfilled two. His overt function might be to supply ribbons, nutmegs, mechanical services or spiritual redemption—traveling preachers were not uncommon, proposing to the busy housewife that they might kneel and pray together—and for this he would almost certainly be welcome. But he would also be welcomed as a provider of news, gossip and fresh human contact. For the first he would get paid; for the second, he might get a meal and a night's lodging.

Had Singer stuck to his machines, he would have been perfectly acceptable and respectable in this society. But it transpired that he did not intend to do so—thus giving the first indication that, even by the standards of his own society, he would not achieve respectability.

2

The Strolling Player

Singer's real dream and intention—and it was one which he never really gave up throughout his life—was to be an actor. But if itinerancy was respectable, acting definitely was not; it was condemned by the clergy and by all morally-minded citizens. As one remarked, "Usually the characters of the actors comport with the scene. Such coarse buffoonery, set off by stale songs and monkey dances, only degrades and corrupts the spectators." On these grounds, theaters were shut and acting suppressed wherever possible. But such opposition naturally served only to sharpen the wits of really determined thespians, and shows of all kinds were given in taverns, bars or indeed any suitable room, often cunningly disguised to defeat moral opposition. Almost anything could, it seemed, be dressed up to look like a tract—as this playbill shows:

KING'S ARMS TAVERN, NEWPORT, RHODE ISLAND
On Monday, June 19, at the Public Room of the above Inn, Will be delivered a series of
MORAL DIALOGUES
In Five Parts,
DEPICTING THE EVIL EFFECTS OF JEALOUSY
AND OTHER BAD PASSIONS,
AND PROVING THAT HAPPINESS CAN ONLY SPRING FROM
THE PURSUIT OF VIRTUE.
Mr. Douglas will represent a noble and magnanimous Moor named Othello, who loves a young lady named Desdemona, and after he

has married her, harbors (as in too many cases) the dreadful passion of jealousy.

> Of jealousy, our being's bane,
> Mark the small cause, and the most dreadful pain.

Mr. Allyn will depict the character of a specious villain, in the regiment of Othello, who is so base as to hate his commander on mere suspicion, and to impose on his best friend. Of such characters, it is to be feared, there are thousands in the world, and the one in question may present to us a salutary warning.

> The man that wrongs his master and his friend,
> What can he come to but a shameful end?

—and so on.

Despite the moralists, a fair variety of theatrical experience was available in towns such as Rochester in the 1820s (although by 1838 a historian of the town was able to give public thanks that "neither theatre nor circus are now to be found in Rochester"). At the crudest and most sensational end of the scale, it was here that "Sam Patch, that notorious fall-jumper, finished his mad career in the autumn of 1829. There are two falls within a short distance of each other, the one descending twelve, the other ninety-seven feet. Upon a projecting rock about the centre of these he erected a scaffold twenty-five feet in height, making together 122 feet, from which he fearlessly leaped into the gulph beneath . . ." He had, it seems, done this several times before with no ill effect; but this time he was drunk, and his body was not found until the following spring at the mouth of the Genesee River. No doubt young Singer was among the eager spectators.

More to his taste, however, was the kind of fare available from the itinerant theater troupes which toured the country—current farce and melodrama and favorite Shakespearean tragedies. One such troupe, led by Edwin Dean, father of the famous actress Julia Dean Hayne, visited Rochester in 1830. Isaac summoned up his courage, approached Dean, and asked if he might join the troupe.

On the whole, such approaches were not encouraged except where it seemed that the newcomer might bring some solid cash with which to back the company. Sol Smith, for example, who went on to become one of the best-known actor-managers of his day, made just such an approach to an itinerant group. He had learned the lead part in *Young Norval*, and hoped that he might be given a chance to play it. In fact he found that it took the utmost persuasion to be allowed to play the waiter in *Raising the Wind*—only to find that on the crucial evening the performance was canceled, and the next morning scenery, theater and actors had all vanished.

Young Isaac, however, appears to have had an easier ride, at least at first. He was a fine-looking youth—over six feet tall, big and blond with a cheery manner—and perhaps that swung things in his favor; the whole of his subsequent career indicates that he must have had considerable charm and magnetism. People were willing to let themselves be carried away by him—sometimes literally—and quite often to their subsequent regret. Here perhaps is a case in point. Dean asked him what were his favorite parts. Singer replied, King Richard III, Macbeth and Othello. He was asked to recite some of Richard's lines, and on the strength of this he was hired for the title role. He rehearsed, said Dean, earnestly. The big night came. When he declaimed the famous lines "A horse! A horse! My kingdom for a horse!" there was a roar of applause from his friends seated in the front—a gratifying reaction which was repeated on several nights following. On the strength of this, he left town to tour with the troupe—and this, perhaps, is the real explanation for the phenomenal brevity of his apprenticeship as a mechanic.

In our own day, the ability of a young man to recite long passages of Shakespeare on request would indicate a certain cultural awareness which somehow does not accord with the other, undoubtedly correct, view of Isaac Singer as an ill-tutored young mechanic from a one-horse frontier town. But the two were not so hard to reconcile then as they might be today. If you could read at all—and Isaac could read, if not particularly well or easily—the most likely available reading material, apart from the Bible and the local newspaper, would probably be a volume of Shakespeare. Many people were familiar with his poetry, and also with that of Milton and Pope. Politicians made literary allusions and quotations in their speeches with no fear of these being missed. Newspaper editors were fond of beginning editorials with a suitable quotation. Added to this, people would be familiar with Shakespeare off the page as well as on. All the touring companies included Shakespeare in their repertory, and many members of their audiences might have acted the parts themselves on the stages to be found in schoolhouses and colleges which were used for amateur theatricals. In a town the size of Rochester there were sure to have been opportunities for such things, of which young Isaac probably took advantage.

Unfortunately it was found that his reception in other towns was less warm than it had been in Rochester, where he obviously had a special local appeal. Although he liked to boast later in life that he had been "one of the best Richards of his day" there were those who, at the time, did not seem to agree with this assessment. One critic described his performances as "crude and bombastical," and considering what is known of his general behavior this was probably a fair description. Singer was never one to speak when he could shout; he would rarely fail to emphasize a point with a suitable swearword. It seems unlikely that he would have made a very subtle Richard. Perhaps the job in Auburn was a welcome stop-gap, taken when Dean could find no more parts for him.

So the pattern of the next several years established itself. He would take an acting job, or indeed any job connected with the theater, when he could; and work as a mechanic when he needed the money. In fact he needed money sooner than might otherwise have been the case because, in December of that eventful year 1830, he got married, thus acquiring responsibility for someone other than himself. The girl in question was named Catharine Maria Haley. She lived with her parents, Henry and Mary, and her three brothers in Palmyra, a small town not far from Rochester. She was fifteen when the marriage took place; Isaac was just nineteen. It was the beginning of an epic amorous career, both in and out of wedlock.

The pertinent question with regard to this marriage is why it should ever have taken place. Certainly it never did Catharine much good; indeed, as things turned out she could scarcely have done worse. From Isaac's point of view it made more sense, at least in the long run: for thirty years, until Catharine finally divorced him in

1860, it effectively shielded him from having to marry anyone else. However, even the greatest cynic could scarcely believe that this is what he had in mind during the actual ceremony. A simpler and more probable explanation is that what he had in mind was getting Catharine into bed. This was his invariable reaction throughout life on meeting a pretty girl. In this as in other walks of life, single-mindedness paid off: he almost always got what he wanted. In this instance, however, the family probably insisted that he marry her first. It is easy to see what a glamorous figure Isaac must have cut before a young girl like Catharine. Not only was he such a fine figure of a young man, but he had just scored a great success in a leading part with a theatrical company. Catharine probably thought herself the luckiest girl in New York State. Her parents may have thought otherwise. At any rate, they made sure that the ceremony was properly organized and witnessed by a respectable number of neighbors. The wedding did not take place in church: it was performed by a local justice of the peace named Smith. Catharine's brother William acted as best man, and a friend named Amanda McMichael was bridesmaid. Catharine's parents were present, as were a number of neighbors; but none of Isaac's relatives seems to have been there. One gets the feeling that the whole occasion was of much more interest to the bride and her party than to the groom.

For a short time the young couple boarded with the Haley family, and then set up house for themselves nearby. At the time of his marriage, Isaac was making a living as a woodturner. Later, he got a job in a dry-goods store at Port Gibson, another small town, and the Singers moved there. It was in Port Gibson that their first son, William, was born in 1834. Isaac seems at this time to have been a classic embodiment of the phrase "jack of all trades, master of none." The truth was that none of the jobs he might get held any real interest for him; they were mere interludes, necessary interruptions to what he considered his real vocation. "Most of his time was spent giving performances," remembered Catharine's brother, Horace Haley, years later. He was rarely home, but traveled about the country taking whatever jobs he could find connected with the theater. If he could not be an actor, he would take a job as advance man for a company but, since most of the troupes were so small, there would generally be a part or two for him to perform as well.

It does not sound the ideal basis for a marriage. Such a way of life would strain a relationship even if both parties were devoted to keeping it going—which Isaac was not. His attitude toward Catharine was that which he was to manifest toward all women with whom he became involved throughout his life: an attitude crude but, if we are to judge by results, effective. Indeed, one of the aspects of Singer which most annoyed his more moral acquaintances was his undeniable success as a womanizer. Singer loved women for their bodies and not for anything as abstract and uninteresting as their minds. He liked, and would always acknowledge and even support to the best of his ability, any children of his they happened to bear. He would not, however, let such trifles, or the mere fact of marriage or long association, interfere with his business or pleasure. He was, no doubt, the kind of man who adds a certain backbone of solidarity to the feminist movement. Men like him were the reason why respectable young girls should always be chaperoned. And yet he was not all bad, as the various women in his life probably repeated on the frequent occasions when they must have asked themselves how they had ever landed themselves in such a situation. He did not tend to abandon them, or disown their children. They simply had to resign themselves to sharing his person and support with others—and there is no denying that at times he did spread himself rather thin.

Many a young wife nurses dark suspicions when her handsome husband spends all his time away from home, ostensibly looking for work; Catharine's, it seems, were justified. In small-town society, gossip travels fast. "His intimacy with the female part of the population was severely commented upon, and much sympathy was expressed for his wife," was how one newspaper put it. (The same source seems to have a curious notion of what it is in men that women find attractive. "His powers of imitation made him attractive to the women," it remarked.)

In 1835 or 1836, the Singer family moved to New York City, where Isaac took a job at Hoe's press shop. At this time it seems that the marriage was still intact despite the gossip, although it had not been without its arguments. However, in the spring of 1836, no doubt to Catharine's fury, he abandoned his job and left the city with a company of strolling players, acting as their advance agent. Catharine remained behind with William.

In the course of their tour the company arrived in Baltimore, where they stayed for some time; and here among the audience one night Isaac noticed a young girl of considerable beauty, with soft brown hair and blue eyes. After the performance he made a point of seeking her out. Her name was Mary Ann Sponsler, and she was eighteen. Her father was in the oyster trade—Baltimore, then as now, being famous for its sea-food—and ran a packing business, employing several men to can the oysters. Singer visited Mary Ann at home, and became friendly with the family; eventually he lodged with them for some weeks. He did not mention that he was married, and the family accepted that he was interested in Mary Ann. Indeed, they encouraged his suit—he seemed an exceptional young man, no doubt, energetic and good-looking—and by the end of his stay he and Mary Ann were betrothed.

Singer now obviously found himself in a tricky personal position. Before he could acquire a new wife, something would have to be done about the old one. He hurried back to New York to try and make some suitable arrangements, extracting meanwhile a promise from Mary Ann that she would follow in September when they would, he assured her, be married.

The first thing he seems to have done on his return was to make Catharine pregnant again: their second child, a daughter, Lillian, was born in 1837. However, their reconciliation could necessarily not last very long. The couple quarreled. "Which was at fault was not known," says one report. "Each accused the other of unfaithfulness, and hard words passed between them." Catharine was a sadder and a wiser woman at twenty-one than she had been at fifteen. However, whether or not she had been unfaithful to Isaac, there seems little doubt which of them must have instigated the quarrels: he had everything to gain from a separation, she had everything to lose. Inevitably, they did indeed separate. Catharine stayed on in New York for a while, but after Lillian was born she returned to her parents in Palmyra—no doubt much to her mortification.

In September Mary Ann arrived, as promised, in New York. The idea of a respectable young girl traveling alone on such a journey would have been unthinkable in Europe, but self-reliance was a virtue in America, and foreign observers all described how ladies traveling alone were treated with great respect. At any rate the Sponslers obviously trusted Mary Ann and, unlike the Haleys, were not suspicious of Isaac. The very existence of such freedom argues, rather paradoxically, that morals were generally rather puritanical, and that most young men could be trusted in such a situation. Be this as it may, Mary Ann arrived alone, Singer welcomed her, and in default of an immediate marriage ceremony invited her to live with him, at which she naturally demurred, reminding him of his promise. There was, of course, very little that he could do about this. From time to time he made excuses, but finally he had to admit—not exactly that he was married, but rather that he was in difficulties with another woman who claimed to be his wife. He explained that he had parted from her and would never live with her again, and that he had very good grounds for divorce. He then begged Mary Ann once more to come and live with him as his wife, promising that as soon as he could, and when he had got some money together, he would take legal steps to relieve himself of his unfortunate married state.

What was Mary Ann to do? She could have turned around and gone straight home again, but that would have involved considerable loss of face—never a small detail at the age of eighteen—and anyway, she was no doubt in love with her glamorous Singer. She wanted to live with him—it was just that she would naturally have preferred to be married first. Perhaps it was agreed that her parents, safely in Baltimore, need never know. Besides, she would be married in all but name. Such arguments, presented by a twenty-five-year-old man of the world to a girl of eighteen, carry some weight; accordingly, she capitulated, and when the couple did return to Baltimore they announced themselves as man and wife. On 27 July 1837, Isaac Augustus Singer, the first child of Isaac and Mary Ann, was born.

* * *

Singer was only seven years older than Mary Ann, but the gap between eighteen and twenty-five—especially a well-traveled and
cynical twenty-five—is a wide one. He called her "Ann," but she
never seems to have called him "Isaac"; when she spoke to him she
called him "Father" and referred to him in public as "Mr. Singer."
In later years, Singer frequently remarked that "By the gods, I don't
know what I should do without Ann," and indeed she was to give him
unfailing support through what were to be very hard times. But
things did not start out that way between them. No sooner was the
baby born than Isaac announced he was once more for the road. It is
possible that he presented Mary Ann with the unenviable choice of
staying with the baby in New York or accompanying him on the
tour, but more probable that the second alternative was never mentioned and that he proposed to treat her as he had treated Catharine.
It was not for this that Mary Ann had sacrificed her chances of
marriage, however, and she did not intend to put up with it. She
picked up the baby and went back to Baltimore, announcing that her
husband had deserted her.

There is no clear record of how Isaac spent the next two years,
but it seems likely that acting jobs, if he found any, were uncertain
and hard to come by. The next time we catch sight of him, in 1839,
he had for the moment given up the struggle and was working as a
laborer in Chicago, where one of his brothers was a contractor for the
new Lockport and Illinois Canal, then being built.

The success of the Erie Canal, which opened up a direct water
route from the Great Lakes to New York, and the enormous prosperity
it had brought to the towns along its route, encouraged a great spate
of canal building in the years following its opening. This was given
even more impetus as steamboats developed and became faster, if not
necessarily safer. Many of these new enterprises were supported by
public funds, but they did not all turn out to be profitable, since
canals were very soon superseded as the country's main transport
network by railroads (which were almost all privately financed).
However, while the canal boom was on, enormous labor forces were
required to perform the mammoth excavations, precursors of the
gangs of navvies who built the railroads across the continent.

The only power available for shifting earth and rocks at that time was muscle power, at best of horses but mostly of men. Here Singer gave the first indication of possible things to come. Such a situation was obviously ripe for the introduction of labor-saving machinery; and in 1839 Singer invented his first machine. It was a machine for drilling rock, "operated . . . by a crank, and its append-ages, in such manner, that the drill is raised the same height at each successive stroke, without altering, or setting, any part of the machinery, from the commencement of the hole to its termination, other than re-moving the drill, for the purpose of clearing the hole." The machine was powered by horses which walked in circles within the wooden framework to turn the crank, and the inventor claimed that the drill bit was such that "the hole will be round, and true . . . avoiding those three cornered holes usually made with a flat drill." The machine was patented the same year. It obviously worked, since Singer was able to sell the patent for $2,000—a considerable sum, and certainly the most money he had ever had in his whole life.

The selling of the patent was an event of some importance for Singer. It was by now quite clear that he did not picture himself maintaining a family in modest prosperity won by a lifetime's honest toil. He was cut out for higher things: he was still hoping to become a famous tragedian, and this was an ambition with which family life could not compete. But even he cannot by this time seriously have expected to get rich by acting. He was now twenty-eight years old; he had been knocking around the fringes of the theatrical world for nine years, and he must have known that even the most famous of the actor-managers rarely made very much money, while most troupes existed in a state of rarely interrupted penury. On the other hand, here now was proof that there was money to be made with inventions, and that he was capable of making it.

It would therefore have been understandable if he had now been bitten by the inventing bug—one which has proved in many cases quite as pernicious as the acting variety. It is a well-known charac-teristic of inventors that each modest success, even if it occurred years before, bolsters their faith in the ultimate blockbuster which is sure to come, just as the memory of one win sustains a gambler through countless losses—or one stage success blots out the memory of the subsequent failures. But even when the days of great success arrived, Singer was never able to transfer his allegiance from acting to inventing. Acting was and remained his passion; he just happened to be more successful at inventing. The time finally came when he was forced to admit this and spend his life concentrating on his inventions, but it was always the histrionic side of these on which he liked to dwell: their promotion, the recounting of the dramatic tale of how they had occurred to him, the personal fame which they brought. He never forgot his acting days and remained a passionate theatergoer. At another level, his personal life always seems to have remained for him in some way a stage which he liked to people with himself in various roles. Inventions brought him the means thus to indulge himself in style, but they remained for him a source of funds and never an end in themselves. He would not have starved for them in a garret, as he was prepared to starve—and let his family starve—for the chance to act.

However, the time for admitting defeat as an actor had not yet come, and the success of his rock drill provided him with the means to indulge his passion once again. The *Atlas*, in 1853, remarked that "he soon scattered the proceeds with the lavish improvidence which so generally characterizes men of genius," thus discreetly omitting to say what he actually did with the money. At this later date Singer seems to have wished to draw a veil over his career as an actor; perhaps the memories were still too near and too painful, or perhaps he had other more cogent reasons for not wishing to be too closely and publicly identified with Isaac Merritt the actor—which is what he now became. He had always used his middle name as his stage name; and now Isaac Merritt was one of the twenty actors listed in Chicago's first city directory, the leader and manager of his own company: the Merritt Players.

Despite the fact that she retired home in a huff in 1837, Isaac seems not to have lost touch with Mary Ann or the Sponsler family, and now that he had money he took the opportunity to bring Mary Ann and young Isaac Augustus, known as Gus, out to Chicago to join him. Her brother Charles Sponsler also came, and these seem to have formed the core of the Merritt Players, which was an essentially family affair. Mary Ann took acting lessons from Isaac in order that she might, as Mrs. Merritt, assume the female roles, and Charles traveled with them for some time as musician and general dramatic assistant. After a while he seems tb have dropped out, leaving Isaac and Mary Ann on their own. Altogether the Merritt Players remained in existence in one form or another for the next five years.

It was a small band, but this did not necessarily limit their repertory too drastically. Most such touring companies were small, but compensated by doubling or tripling or even quadrupling up on roles. Sol Smith, for example, toured in a company of four men and two women, which presented *Pizarro*, a drama with a total cast of seventeen.

Sam Drake (as Pizarro), after planning an attack on the unoffending Peruvians while engaged in worship at their ungodly altars, and assigning his generals (me) their several posts, in the next act is seen (as Ataliba) leading the Indian warriors to battle. He is victorious, and goes to offer up thanks to the gods therefor—when, presto! on comes the same man again (as Pizarro) smarting under the stings of defeat! Fisher (as Las Casas) calls down a curse on the heads of the Spaniards; throws off his cloak, drops his cross, doffs his grey wig, and appears in the next scene as the gallant Rolla, inciting his brave associates to deeds of valour. Alexander Drake, as Orozembo, gives in the first scene an excellent character to the youth Alonzo, proclaiming him to be a "nation's benefactor"; he is then stuck under the fifth rib by a Spanish soldier (that's me, again), and is carried off by his slayer; he then slips off his shirt and skull-cap, claps on a touch of red paint, and behold! in the next scene, he is the blooming Alonzo, engaged in a quiet tête-à-tête with his Indian spouse. For my own part, I was the Spanish army entire! but my services were not confined to that part. Between whiles I had to officiate as High Priest of the Sun; then, as Blind Man, feel my way, guided by a little boy (one of the Fisher children) through the heat of battle to tell the audience what was going on behind the scenes; afterward, my black cloak was dropped, I was placed as sentinel over Alonzo. But, being a novice, all my exertions were as nothing in comparison with those of the Drakes, particularly

Sam, who frequently played three or four parts in one play, and after being killed in the last scene, was obliged to fall far enough off-stage to play the slow music as the curtain descended.

It is unlikely that *Pizarro* would have been part of the Merritt Players' repertory since they confined themselves to Shakespeare and temperance dramas, which they performed in churches.

The temperance movement was at this time very strong throughout the United States—a strength matched only by the roaring trade conducted in taverns. In 1832, it was estimated that seventy-two million gallons of distilled spirits were consumed by Americans each year—six gallons for every man, woman and child in the country. Apart from the moral and physical depredations, concern was expressed at the loss (through idleness of drunkards, support of paupers, criminals, etc.) of, as one observer estimated, $94,525,000 annually—without taking into account loss from shipwrecks, fires caused by intemperance, costs of litigation, and other such mishaps. There were thus solid financial as well as moral reasons for supporting the temperance movement.

The movement had various propaganda outlets. Numerous tracts were produced and sold; the famous Parson Weems—he was the man who invented the story of young George Washington and the cherry tree—would invade barrooms, delivering a diatribe on the evils of liquor and drunkenness and then selling—apparently with great success—copies of his *Drunkard's Looking Glass* at 25 cents each. Stirring orations were delivered to temperance societies, sometimes by reformed drunkards who recited their misadventures due to strong drink, a method that was found particularly effective. Nothing, however, more vividly conveyed the horrors of drunkenness than their portrayal, with no ghastly detail omitted, upon the stage. "The dreadful power of strong drink was never so thoroughly impressed upon me as when I sat, one afternoon . . . and waited for the successive risings of the curtain, with its picture of the Bay of Naples, which separated us from the stage," recounted one enthralled member of a temperance audience.

As the grog shop was revealed in all its horror, we felt that dramatic realization was in our midst. When the drink-crazed father hurled the rum bottle into the *left wing,* the little daughter obediently trotted on from the *right* exclaiming: "Oh, Papa, you have killed me!" Then she fell dead in the center of the stage, and we suspected, as I have often suspected since, that the Demon Rum does not often get fair treatment from his foes. We felt sure that not even a drunkard could pitch such a curve as that.

It is quite easy to imagine that, quite apart from any moral considerations, such scenes provided roles greatly to the liking of the always melodramatically inclined Singer. (They may have been morally to his liking, too: he did not drink—an unexpected and unexplained characteristic in one generally so intemperate.) Not even his favorite *Richard III* provided more scope for pathos. But irresistible as such scenes might often be, they were not enough to provide Isaac and his growing family with an adequate living.

The life of a traveling actor at this time was a very hard one. All who describe this life agree on the extreme difficulty of raising any kind of an audience outside the big cities. Harry Watkins, who traveled with a troupe consisting of himself and two others in the 1840s, recorded that on the night of his benefit show (the evening when he would receive half the takings) the ticket window took only two dollars. "There were two men in the boxes, one boy in the pit, and a Nigger woman in the gallery," he recalled. The famous actor-manager John Bernard found himself and his troupe performing one evening to an audience of about twenty, with

more than double that number standing outside and contenting themselves with reading the bill, catching fragments of the songs, and imagining the rest. At night, therefore, some of the company resolved on going round the town to give all the pretty girls a serenade, flattering themselves that if the women—the most influential part of all communities—once heard their voices, the next evening our room would be as packed as the "Black Hole." The compliment was so novel in the quiet streets of this secluded place that every window flew up and some score of female faces popped out, which was considered a favourable omen. But the Delaware maidens were better calculators; the next night we performed to only ten dollars; and I learned that "they could not think of *paying* to hear what they had already heard for nothing."

This lack of popularity was often attributed to opposition from the clergy, who indeed did contrive to shut down several theaters at one time or another; but Frances Trollope, observing the same phenomenon, did not believe this. "The cause, I think, is the character of the people," she said.

> I never saw a population so totally divested of gayety. They have no fetes, no fairs, no merrimaking, no music in the streets, no Punch, no puppetshows. If they see a comedy or a farce, they may laugh at it; but they can do very well without it; and the consciousness of the number of cents that must be paid to enter a theatre, I am very sure turns more steps from its door than any religious feeling.

In addition to lack of enthusiasm, there were other hazards a traveling actor might have to face. Sol Smith recalled the eagerness of local sheriffs to serve writs against his company for playing without a license, which might mean a forty-dollar fine to be paid out of only fifteen dollars in receipts. His company was once stopped by the local constables on suspicion of being "pirates." When asked the reason for such a wild accusation, they replied: "That's a good one. In the first place, what can *honest* people do with such a heap of plunder as you are toting in that wagon? Nextly, your confessions last night before Peggy Duncan, while you were eating supper. Didn't one of your men like to have been taken before he escaped to the ship, after killing a Don?" Smith concluded that nobody could succeed in this life without an equable temperament and considerable wit.

Even more outlandish difficulties might beset the travelers. Smith described how at one point his company was floating down the Allegheny River in two boats. One went ahead to put up a white flag on the bank when a suitable town was reached. The second followed, and after some time a white handkerchief attached to a pole was made out on the bank, but there were no houses to be seen. They wondered what this could mean. Finally a voice called from the bank: "Oh, you're looking for the houses. Bless ye, they are not built up yet; but we shall have some splendid buildings shortly!"

At the end of all this tribulation, there was of course no guarantee of making a living. "Had supper this evening," noted Harry Watkins, "the finest meal for some time as we have been living on dry bread for a week."

It was a note that Singer and his little band might have made more than once in the course of their acting career. They experienced the direst poverty during these years. With no real home, they wandered around the country carrying all their possessions in a wagon. Even this had occasionally to be pawned in order to get enough money to buy food. Meanwhile the children kept coming. In 1840, Voulettie Theresa was born; in 1843, John Albert; in 1844, Fannie Elizabeth. "I am the happiest man in the world," said Singer. "I have boys and girls alternately." But although he was fond of his children, especially the girls, he cannot always have felt so sanguine, and nor could Mary Ann. "Ann is a good woman and a faithful wife," he would observe. As to the first, she certainly seems to have been so; as to the second, it is unlikely she ever had time or energy to consider anything else.

She may also have been unwilling to try his temper too far, for the hard life and lack of success did not improve his disposition. Many of the people who met Singer remarked that he was rough and violent in his manner and tended to intimidate all who came into contact with him, including his family. By 1843 this characteristic was already marked enough to be remembered by one of the hotel-keepers with whom they stayed in the course of their wanderings, a Mr. Tuttle. "In the year 1843 or 1844," he wrote some years later,

I kept a hotel in the town of Pequa, Ohio. Singer, calling himself Merritt, with his wife and two children, came to my hotel and remained nearly a week, giving recitations, mostly from Shakespeare's plays; they giving their entertainments in the ball room of the hotel. I remember my wife gave up our family room to the "distinguished elocutionist" and his family while they remained. My recollection of Singer is that at the time he was very poor in pocket, shabby in person, and disposed to be rough and unkind in his manner. Mrs. Singer was quite the opposite; refined, amiable and courteous. The children, a boy and a girl, were bright, well mannered children. They all left my hotel in a common two horse wagon, and their whole baggage would not have brought $10. I gave him $3 to pay his way to the next town.

Such a life could not be endured indefinitely. In 1844, the Merritt Players finally gave up the ghost, leaving Isaac, Mary Ann, Gus, Voulettie, John and Fannie more or less stranded in the village of Fredericksburg, Ohio. Isaac may have been reluctant to take this step off the stage, but it is doubtful if Mary Ann felt anything but relief. If they were not affluent, at least they were settled in one place.

The reason why they stopped just at this time and place is unclear. Perhaps they were influenced by the fact that a suitable job happened to be available there for Isaac, in a printshop that did a lot of posters and advertisements. Such items were mostly set up in wood block type; it was his job to carve the type as it was needed. It was steady work of a kind at which he had always been good, and Fredericksburg was a pleasant, quiet, rural sort of a place. No doubt Mary Ann was happy enough; but not Isaac. Such factors had never weighed heavily with him. His destiny, as he saw it, had never been to molder in an obscure job in an obscure town. However, it was by now obvious even to him that he was unlikely ever to make an enduring name, or even a living, as an actor.

Thus at the age of thirty-three—an age at which most people even today, and certainly then, have a fair notion of how they are probably going to spend their lives—Singer had totally to reconsider his. He had spent the previous fourteen years, the best of most people's lives, trying hopelessly and persistently to do something at which he had ultimately failed. This is never easy to admit, and especially not for a character like his: ambitious, exhibitionist, swaggering, deeply conscious of the impression he made. That he over-reacted to success later is perhaps not surprising in view of the fiasco which crowned all his early efforts to tread the path to glory.

The alternative immediately presenting itself, the steady type-carving job in Fredericksburg or its equivalent, was a humiliating one. He had not spent fourteen years traveling around the country and often suffering acute poverty and discomfort simply to return to what he had rejected in the first place—the quiet life of a mechanic out in the sticks. There remained only one other, more satisfactory alternative. He had made good money before, and quickly, by turning his mechanical skill to invention. The circumstances of his present job suggested a means whereby he might achieve the same success again. He began work on a machine which would carve wooden type.

He seems actually to have brought this machine into operation in Fredericksburg, since the *Atlas* reported nine years later that "a large quantity of type was made by this machine, which were extensively sold throughout the country, the quality of the work being very superior. Becoming dissatisfied with the management of his co-partners at Fredericksburg, he abandoned the business there . . ." It is questionable, however, whether that last sentence really conveys what happened. For one thing, it is most unlikely that Singer was ever a co-partner in the Fredericksburg printing business. He might well have liked to imply later that he was, just as he preferred to draw a veil over the dispiriting years of touring with the Merritt Players with those phrases about "scattering the proceeds" of his first invention with "the improvidence of genius." When this interview was given he was trying to establish himself as a successful New York businessman, and he naturally wished to bring his past up to scratch. However, a partnership usually requires some sort of monetary investment—and Singer certainly had no money in 1844 or 1845. There is, on the other hand, no question that he abandoned Fredericksburg. The local recollection is that one day he simply disappeared into the blue—went out on some errand and never came back.

His next port of call was Pittsburgh, which was the nearest large town to Fredericksburg. Here he established himself and his family (who presumably followed when he sent for them) in 1846. Continuing in his new métier, he set up a workshop for making wood type and raised sign letters, using his undoubted mechanical skill to set up and arrange all the necessary machinery himself. This business proved quite successful, and the Singers (as once again they were all known) remained in Pittsburgh for the next three years. Meanwhile Isaac was working on his type-carving machine and improving it in various ways, finally patenting it as a "Machine for carving wood and metal" on 10 April 1849. Once again the Singers packed up and set out on the road. This time their destination was New York, where Isaac hoped to conquer the world with his machine.

3

The Sewing Machine

The way we view a person's characteristics can be transformed by consciousness of his subsequent success or failure. Until now Isaac Merritt Singer's life had been, if not a failure, very far from a success, and it seemed unlikely that he would ever hit on an idea which would make him rich. He was now thirty-eight years old, with two wives and eight children (Jasper Hamet and Mary Olive having been added to the brood while they were living in Pittsburgh). All he possessed was his type-carving machine, which was indeed an ingenious one: the motion of the cutter was controlled by a system of "pantograph levers" provided with a pointer or tracer, so that essentially it was a machine which would carve as you drew. But it had two defects: no commercial prototype had yet been built, and, worse, the age of wooden type was coming to an end so that, however ingenious it was, this machine was unlikely to make a fortune for anybody.

Most of the men who made the kind of money which did eventually come Singer's way were well on the road to establishing themselves, if they had not already done so, by the time they reached their late thirties. The fertility of invention, energy and perseverance which is so necessary for self-made men is also, on the whole, a characteristic of youth. The time for penniless casting about was logically at least ten if not fifteen years past. A young man with no capital but a stock of good ideas is a figure to encourage; a middle-aged man in the same position is more usually pitiable. Many must have been the gloomy voices which pointed out that only a man as blustering and irresponsible as Singer could have landed himself and his family in such a hopeless situation.

On the other hand it might be said that Singer's career to date showed a character of exceptional vigor and optimism. His ambition had not been quenched or even dampened by past misfortunes. He could easily enough have settled down to a humdrum but safe life in Fredericksburg or Pittsburgh, but he would not do so; he was still ready to risk everything on the final throw. Annoying as such an attitude may well have been to the long-suffering Mary Ann, it was probably, and paradoxically, this reckless optimism which was one of Singer's most attractive qualities.

It was certainly a characteristic very much in tune with the times. Vigor and optimism were in the air. If, today, staking everything on an invention seems just about as crazy as blowing a fortune to set up a company of strolling players, the same was not necessarily true then. America in the mid-nineteenth century *was* the land of opportunity: and this was particularly true for inventive tinkerers such as Singer. For if there were opportunities for men in America, there were even greater opportunities for machines.

It is perhaps not generally realized how long-established are American and European attitudes toward labor-saving machinery, or the lack of it. The American visitor to Europe has traditionally expected to find unheated houses, antiquated household equipment with little labor-saving gadgetry, and on the industrial side an irritating lack of awareness of productivity and timesaving. Such expectations, even today, are often fulfilled. The European visitor to the United States, on the other hand, is constantly amazed by the automation permeating all aspects of life, some genuinely making life easier, some apparently introduced more for sheer pleasure in the machine itself.

Nearly all the European visitors who flocked across the Atlantic in the nineteenth century to see what life in a new world was really like noted that this state of affairs was already very much in evidence. As early as 1820, the political economist Friedrich List reported: "Everything new is quickly introduced here, and all the latest inventions. There is no clinging to old ways, the moment an American hears the word 'innovation' he pricks up his ears." In the decade 1820–30, patents in the United States averaged about 535 annually, while in Britain, the most advanced industrial nation in the world, the annual average was only 145. (It is, however, possible that these figures were affected by the lax state of the patent laws in the United States. While many Americans did their best to satisfy the national longing for innovation, others, equally astute, capitalized on another aspect of it. Before 1836, the American patent laws were such that anyone claiming to be an inventor was allowed to patent any device provided he could prove that it was not actually harmful to the community. In this way, many imaginative and unscrupulous people made a lot of money claiming monopoly rights and royalties on devices already in use, under threat of suit for infringement. After 1836, this law was tightened up, a fact which was subsequently to cause Singer a considerable amount of trouble.)

The reasons for this welcoming attitude to novelty and invention are easy to find. Many new Americans had made the journey from a Europe overstocked with cheap, unskilled labor (and with a well-entrenched class of jealous traditional craftsmen) to the new continent where labor of any sort was scarce, and skilled labor scarcest of all. Land was cheap, and wages were high. The result was that any hard-working man in the United States could reasonably expect to become his own master within a very few years. Almost all Americans (as some European visitors noted with disgust) regarded working for another person as a degrading passing phase, to be discontinued as soon as circumstances allowed.★

In such a situation, any machine which could save labor was

★ This attitude shocked Frances Trollope, for instance, to the core: "The greatest difficulty in organising a family establishment in Ohio is getting servants, or, as it is there called, 'getting help,' for it is more than petty treason to the Republic, to call a free citizen a *servant*. The whole class of young women, whose bread depends upon their labour, are taught to believe that the most abject poverty is preferable to domestic service. Hundreds of half-naked girls work in the paper mills, or in any other manufactory, for less than half the wages they would receive in service; but they think their equality is compromised by the latter, and nothing but the wish to obtain some particular article of finery will ever induce them to submit to it."

naturally welcomed—as Singer had already found ten years earlier. In the process, the standards of absolute excellence so important to many Europeans might have to go by the board: the great thing was to get the work done. "All grain is here cut by machine," wrote an Iowa farmer in 1857. "Cradles are out of the question . . . If grain is too badly lodged to be so gathered it is quietly left alone . . . This work done by machinery is not very much cheaper than it could be done by hand, but the great question is—where are the hands to come from?" The mechanical reaper was invented by a young Virginia farm boy, Cyrus McCormick, who was trying to find a solution to the problems faced by his father on their farm. He went on to make millions of dollars, and was an example to every aspiring mechanic: he typified the opportunities that were there for anyone who could hit on a truly labor-saving machine. Peasant communities, where every family over-cultivated the field, might value a hand-crafted finish: in the United States, there was no time to be a peasant.

Besides, and quite apart from the practicality of the thing, the psychology of new invention was attractive to the inhabitants of the young republic. The United States had been founded to get away from the old, stifling ways: what could be more desirable than to create new American ways of doing things? Caution, under these circumstances, was not necessarily a virtue. "Restlessness of character seems to me to be one of the distinctive traits of this people," wrote de Tocqueville in 1831.

> The American is devoured by the longing to make his fortune; it is the unique passion of his life; he has . . . no inveterate habits, no spirit of routine; he is the daily witness of the swiftest changes of fortune, and is less afraid than any other inhabitant of the globe to risk what he has gained in the hope of a better future, for he knows that he can without trouble create new resources again . . . Everybody here wants to grow rich and rise in the world, and there is no one but believes in his power to succeed in that.

It is the picture of Isaac Singer to the life as, clutching the patent papers in which all his hopes lay, and accompanied by the faithful Mary Ann and their numerous family, he made his way to New York in search of a backer.

The first essential was to produce a working prototype; and Singer was not well placed to do this. Many of the inventors of the time were like him in that they were self-taught mechanics tinkering with improvements to the tools of their trade. But most of them would have had access through their job to work space and materials. Singer had neither. What he had to find was someone who would provide him with some space and put up some money so that he could begin the laborious process of constructing his model without letting his family starve.

At first it seemed as though he was in luck. A. B. Taylor and Company, a machine-manufacturing firm with premises in Hague Street, agreed to advance him money and let him have a room in which to build his model. He established the family in rooms on the Lower East Side at 130 East 27th Street, near Third Avenue (not far from where Catharine was then also living, having returned to New York), and got on with building his machine. However, after a short time Taylor's was struck by one of the occupational hazards of the new machine age which it was doing its best to usher in. On 4 February 1850, a boiler burst in the building, killing sixty-three people, injuring as many more, and incidentally wrecking Singer's machine, which he had just completed. (Luckily he himself was not there at the time.) It was the worst such accident ever known; but it was by no means unique—incidents of this sort were occurring almost daily, and were reported in gory detail by the newspapers. Steam power was a boon, but until the art of boilermaking caught up with the general desire for maximum power, it regularly exacted its toll.★

★ This was particularly true of the steam-powered riverboats, which demonstrated the country's almost reckless desire for modernism in trade and communications at (it seemed) almost any cost. These boats were a most dangerous form of transport. The mortality rate from boiler explosions aboard was enormous—fatal accidents were always being reported, and on the Mississippi the chances of an accident on board a boat were about 50–50. But despite this they were easily the most popular form of getting about the country until the railways began to take over in the 1840s (probably because the turnpike roads were so few and appalling and stagecoach journeys slow and exceedingly uncomfortable. What was safety compared to the snail's pace of a coach?).

When Singer set up in business for himself in New York a few years later, his own factory was wrecked by a similar explosion. On this first occasion, it must have seemed like the end of all his hopes. Taylor would advance no more money. The model was destroyed. What was to be done?

It was lucky for Singer and his family that his charm was not reserved solely for women, for now, when he really needed friends, a good friend turned up. This was George B. Zieber, who had owned a bookselling and publishing business in Philadelphia, but who was now conducting generally unsuccessful book jobbing ventures in New York. Toward the end of 1849, just before the disastrous explosion took place, he had been to see Singer's machine, then just completed, which was on display at Taylor's machine shop. There he naturally made the acquaintance of the inventor. After the explosion Singer contacted him, hoping that Zieber, who had already shown an interest in the original machine, might be induced to help build another. He was not disappointed. Zieber was anxious to get back into business, and persuaded some friends of his, Stringer and Townsend, who were publishers with premises on Broadway, to lend him $1,700 in cash and acceptances. With this in hand, he proposed an agreement with Singer to set up one of his machines on a large scale in Boston.

Singer at that time was wretchedly poor [Zieber wrote later in a long, unpublished statement], out at the elbows, without money or credit, with a large family to support. He lived in two small rooms on the second floor of a house in Third Avenue and his children ran about the streets in patched garments. I met him in the street one morning about 12 o'clock, shortly after I had made his acquaintance, when he told me he had not yet had any breakfast, and asked me to lend him half a dollar, upon which I accompanied him to the restaurant under French Hotel.

After I became interested in the carving machine, I was an occasional visitor at Singer's humble home, where dinner and supper were taken together upon stewed meat and potatoes. Forks with two prongs were then used, and we helped ourselves with pewter spoons from one common dish in the center of the pinewood table. But I respected Singer in his poverty. We were "fighting the battle of life" together. He, to get a start in the world, and I, to gain money to pay my debts. We often sat for hours together speculating upon our future fortunes.

Early in the spring of 1850, the proposed agreement was made. Altogether, Singer was to receive $3,000. Zieber was to pay him $600 at once, and he was to receive the balance out of sales of the machines and rights for the State of Massachusetts. Another room was found in the factory of Woorall Bros. in Worth Street, and here Singer was to build another prototype, paid for by Zieber. Most of the $600 went quickly in paying unpaid bills. Some weeks later the Singer family was once again penniless, and the good-hearted Zieber gave Singer another fifty dollars. On 1 June the carving machine was finished, and Singer and Zieber set off with it to Boston.

Zieber had chosen to go to Boston and had bought the Massachusetts rights to the machine because Boston was then the center of the book manufacturing trade, so that high sales could be expected there. But Boston in 1850 was more than just the center of book making: it was one of the most important manufacturing centers in the United States. Agriculture in New England had never been more than marginally profitable. The opening up of the immensely fertile western territories, and the improvement of canal and rail communications for the transport of surpluses, meant that there was no longer the need for each area—indeed, each small community —to be as nearly self-sufficient as possible. The demand for manufactured goods increased and, with it, the development of New England industry.

But New England had other advantages. It had a large population (swelled after the mid-1840s by the constant arrival of immigrants) and a source of free power along the "fall line" of the rivers. The scale was therefore weighted very early in favor of industry, and the establishment of factories, such as the model institutions at Lowell, tipped the balance even further. It became the accepted, respectable thing for girls to leave home and go to work in factories which, in Lowell at least, undertook to house them and provide for their moral, religious and educational welfare.★ The establishment of the factories further tipped the balance against agriculture: farmers complained that the factories attracted labor and raised wages so that hiring agricultural labor became prohibitively expensive.

By 1850, two main types of industry were located in New England. One was the manufacture of textiles and shoes, which had always been traditional to the area, and which thrived with the expanding national market. The other was anything that called for skilled workmanship and intricate machinery. Heavy industry was no longer located here, but had moved, with the discovery of coal and iron, to centers such as Pittsburgh. But the efficiency of machinery and mass production in the New England factories was a source of wonder to European visitors, who saw automatic processes being applied to objects which required skilled craftsmanship at home. By 1850, Aaron Dennison of Waltham, Mass., was already applying mass production methods to such a fine craft as watch-making, which was the province of skilled hand workers in Europe. "They do far more with machinery in all kinds of trades than you do," wrote an immigrant from Sheffield in 1865. "Men never learn to do a knife through, as they do in Sheffield. The knives go through thirty or forty hands . . . If a Yankee can resin a knife, they call him a cutler; and by doing one thing all the time they become very expert and they make some very good knives . . . such patterns as is done easiest by machinery." As we shall see, the years 1850–65 were particularly important in the development of mass production processes, and, in

---

★ This was no great boon for the poor girls. Their working day consisted of 12½ or 13½ hours, and their weekly wage was about $2.50. A bed in a company house was given free, and about $1.25 deducted each week for board. The *Boston Associate* announced that "the company will not employ anyone who is habitually absent from public worship on the Sabbath or known to be guilty of immorality." The matrons of the working girls' dormitories were responsible for enforcing a ten-o'clock curfew and keeping their homes puritanically in order. In 1840, Orestes Brownson said of the Lowell factory girls: "The great mass wear out their health, spirits and morale without becoming one whit better off than when they commenced labour. The bills of mortality in the factory villages are not striking, we admit, for the poor girls, when they can toil no longer, go home to die."

general, if you wanted to build and sell complex machines, Boston was the obvious place to go.

Singer and Zieber rented space at a steam-powered machine shop owned by Orson C. Phelps at 19 Harvard Place, Boston. Here they set up their carving machine and waited for customers to arrive. Over the course of the next few months several people did indeed come to see their machine, but nobody seemed actually to want to order one. It became increasingly apparent that Zieber's money and Singer's time and effort had been wasted. The type carver was destined to be one of the many ingenious inventions which made nobody a profit, let alone a fortune. What was to be done? Singer had no resources and a clamoring family, and Zieber had sunk all his funds into the invention. One can imagine the two men hanging glumly around the machine shop, wondering what they were to do, not wanting to stray too far away for fear of losing a potential customer, although it became more and more obvious that nothing was going to come of the enterprise.

Singer and Zieber had rented a room on the ground floor of Phelps's shop. Meanwhile, in the owner's own main workroom on the next floor, a flourishing business was being conducted; for Phelps was at this time busily engaged in the construction of some of the newfangled sewing machines.

The first serious attempt to market the sewing machine had been made by Elias Howe, who invented such a machine and patented it in 1846. For a variety of reasons (examined in more detail later) it had not caught on, and Howe at this point was in the depths of penury and despair. Nevertheless, a number of other inventive mechanics had by 1850 attempted their own modified version of the machine. Phelps had been familiar with several of these experiments, beginning with those of Howe himself, who had first come to him in 1845. At that time Phelps was a journeyman engaged in making scientific instruments at a shop in School Street, Boston. Howe had approached him in the shop and asked for help in "getting up a machine." He would not divulge the nature of this mysterious machine, but Phelps, who (it would seem) did not take on the job, was afterward told that it was a sewing machine. By 1850, Phelps was manufacturing a machine patented the previous year by J. H. Lerow and S. C. Blodgett and known, naturally enough, as the Lerow and Blodgett machine. It produced a lockstitch using two threads, much as sewing machines do today. However, it looked rather different, since it had a curved, eye-pointed needle, which produced a bowstring-type loop with its taut thread when it penetrated the cloth, while the shuttle with the second thread, below the cloth, instead of moving back and forth as in today's machines, had a circular motion.

It was a clumsy-looking design, but this of course would not have mattered if the machine had worked. However, on the whole it did not. "Of a hundred and twenty completed machines," said Zieber, "only eight or nine worked well enough to use in the tailor's workrooms which occupied the top floor of the building, and Phelps was constantly being called on to repair them." And even when they did work, they were hardly convenient to use. Mrs. Phelps had one of the machines which her husband had built for her own use. "The machine was round," she remembered later, "and the work required to be put upon the outside with pins, which was a great annoyance." The pins pricked her fingers, she said.

It was obvious to Phelps that the machine left room for improvement, and also that there was a considerable future for a more reliable and adaptable sewing machine. However, the necessary modifications would require both inventiveness and capital. It seemed to Phelps at this time that Singer and Zieber had both. The complex and ingenious carving machine was evidence of the one; and Singer had assured him that Zieber had plenty of the other. "He said," asserted Phelps ten years later, "that his friend, Mr. George B. Zieber, had something like eighty thousand dollars to spend in mechanical business, if he could make money out of it." This of course was a gross exaggeration, but if Singer did say this, or something like it, Phelps had no reason to disbelieve it. Phelps therefore tried to interest Singer in his sewing machines.

I went into his carving-room, and Mr. Singer was sitting on a pile of boards near the carving machine, which he had purchased for the purpose of illustrating the movements and the cutting of the machine. I thought he appeared to be most dejected; he had been there some time and did not appear to have much success; and as I naturally wanted to encourage everybody all I could, I said to him, "Mr. Singer, I propose one thing: Leave this carving machine, and go with me into the sewing machine!" "Good God!" said he. "Phelps! do you think I would leave this ponderous machine and go to work upon a little contemptible sewing machine?" Said I, "Mr. Singer, you may be mistaken in your ideas. You have got a machine which you admit cost you two thousand dollars, and you never can sell one for more than three thousand, at any rate, and I think three or four will do the work of the whole Union!" Mr. Singer made answer like this. Said he, "Phelps! There is reason in that!"

It seems unlikely that Singer came around as quickly as that. Phelps was remembering all this ten years after the event, and his testimony was not unbiased, since at the time he was trying to prove that he, not Singer, had been the real inventor of the Singer sewing machine. In fact Singer at first told Zieber that he was disinclined to have anything to do with such a "paltry business." "What a devilish machine!" was another of his comments. "You want to do away with the only thing that keeps women quiet, their sewing!" Certainly humanitarian considerations never swayed his calculations, then or later; Singer would probably have agreed with Stokely Carmichael's remark that the position for women in his scheme of things was "prone."★

However, financial considerations, for a man in Singer's position, tend to outweigh principles, and on reflection Singer decided he might be interested in sewing machines after all. "I don't give a damn for the invention," as he frankly put it. "The dimes are what I am after." His account of the beginning of the affair (given in patent suit testimony some years later) was that Phelps showed him one of the sewing machines and said that, if it could be improved so as to do a greater variety of work, "it would be a good thing." If Singer could accomplish this, said Phelps, he would stand to make a good deal more money from sewing than from carving machines. Singer immedi-

★ What Carmichael presumably meant, however, was "supine."

ately pointed out the main fault of the Lerow and Blodgett machine, which was that the circular movement of the shuttle took the twist out of the thread, making it very weak and liable to break easily.

"And how would you correct it?" asked Phelps.

"In place of the shuttle going round in a circle," replied Singer, "I would have it move to and fro in a straight line, and in place of the needle bar pushing a curved needle horizontally, I would have a straight needle moving up and down." This would, he thought, successfully correct the machine's other main fault, which was that the curved needle was very brittle and always breaking; a straight needle would be far stronger. Phelps later claimed that he, and not Singer, had been the first one to think of both these improvements. However, this seems unlikely; if it was true, why had he not just gone ahead and experimented with the ideas on his own? The sensible thing in that case would have been to approach Zieber directly, since all that was necessary would have been financial, not mechanical, help.

It was Singer who now approached Zieber with the new idea. According to Zieber,

Singer talked with me about supplying money to make an experiment but I felt loth to advance anything out of the small amount yet remaining in my possession, to make experiments in sewing machines, after having just sunk between $3,000 and $4,000 (in the unfortunate carving machine), obtained with difficulty from my friends, whenever I could get it, and I became very much disheartened. Phelps was poor, owed nearly as much as his shop was worth, and required all he could make to pay his current expenses, and as for Singer's being able to obtain, at the time, only $5 or $10, appeared to be entirely impossible. It is well known how difficult it is for poor inventors to obtain money to aid them in making experiments. People, generally, would much rather invest in lottery tickets.

In anticipation, however, that the new enterprise suggested might amount to something, I continued to give Singer $10 per week to send to his family in New York, and paid his own expenses in Boston [Singer and Zieber shared a small room at Wilde's Hotel], until we could see whether there was a chance to make anything at the sewing machine.

Singer meanwhile had prepared a sketch of the machine he had in mind, and assured Zieber that the project "would make us all millionaires." Everyone agreed that the sketch looked promising. Accordingly, a contract was drawn up between Singer, Phelps and Zieber. "I drew up the agreement myself," said Zieber. "It was a plain one, and sufficient to secure to each the interest to which he was entitled, had all the other parties been honorably disposed." (This was a reference to Singer's later behavior.)

The agreement was clear and simple:

Made and concluded upon this eighteenth day of September, One Thousand Eight Hundred and Fifty, between George B. Zieber and Isaac M. Singer of the City of New York and Orson C. Phelps of the City of Boston.

In the first place, the parties above named, have agreed, and by these presents so agree, to become co-partners together in an Improved Sewing Machine, to be called the Jenny Lind Sewing Machine, and to apply for a patent for the same.

And it is further agreed by the said G. B. Zieber, that for the purpose of making an experimental machine, he will furnish the sum of Forty Dollars, and if said machine is successful, whatever further amount is necessary to procure a patent for the same. If the experiment fails, said Forty Dollars shall be lost to G. B. Zieber. He, the said G. B. Zieber, also agrees to attend to the business of the said co-partnership, to assist in making sales of rights and machines, and to do whatever may be a mutual advantage to the whole.

Each party to this agreement shall have power to sell rights and machines which sales shall be valid and binding as soon as the payment is received and shall be divided equally among the parties to said agreement.

Said Isaac M. Singer further agrees to contribute his inventive genius towards arranging a complete machine, and to do everything in his power towards perfecting the work, and all improvements which he, the said Singer, may make on said sewing machine, during the existence of the contract shall belong to and be the property of, the said parties thereto.

Said Orson C. Phelps further agrees, for the sum of Forty Dollars before named, to employ his best mechanical skill in completing a Machine, which machine, when completed and successful, shall be sent as a model to the patent office in Washington, upon which to obtain a patent, and it is also agreed by said Phelps that he will not at any time during the existence of this partnership, engage in the manufacture or invention of any sewing machine, except the one belonging to the parties to this agreement. And it is also agreed that said Phelps shall have the exclusive right to manufacture all the machines sold by the parties to this agreement, at the same prices that could be charged by any other respectable manufacturer.

It is also agreed that the patent shall be applied for and taken out in the name of Isaac M. Singer and Orson C. Phelps—but that said patent shall be the equal property of the three partners to this agreement, each owning one-third thereof.

The contract was signed by Phelps, Singer and Zieber, and witnessed by James Baker, Jr. Its wording seems to make it quite clear where, at that time, the innovations were considered to have originated, despite what might be said later.

Now it was all up to Singer. Once again the prospect of striking it rich gleamed before his eyes. What was more, he knew it was his last chance of obtaining financial support from this source, at least, since the forty dollars represented the end of Zieber's resources. The urgency he felt is clear in his description of the ensuing fortnight:

I worked day and night, sleeping but three or four hours out of the twenty-four and eating generally but once a day, as I knew I must get a machine made for forty dollars, or not get it at all. The machine was completed the night of the eleventh day from the day it was commenced. About nine o'clock that evening, we got the parts of the machine together, and commenced trying it. The first attempt to sew was unsuccessful; and the workers, who were tired out with almost unremitting work, left me, one by one, intimating that it was a failure. I continued trying the machine, with Zieber to hold the lamp for me, but in the nervous condition to which I had been reduced by incessant work and anxiety, was unsuccessful in getting the machine to sew tight stitches. Sick at heart, about midnight I started with Zieber to the hotel where I boarded. Upon the way, we sat down on a pile of boards, and Zieber asked me if I had noticed that the loose loops of thread on the upper side of the cloth came from the needle. It then flashed upon me that I had forgotten to adjust the tension upon the needle thread. Zieber and I went back to the shop. I adjusted the tension, tried the machine, and sewed five stitches perfectly, when the thread broke. The perfection of those stitches satisfied me that the machine was a success, and I stopped work, went back to the hotel, and had a sound sleep. By three o'clock the next day I had the machine finished, and started with it to New York, where I employed Mr. Charles M. Keller to get out a patent for it.

The histrionic qualities of this tale seem almost too good to be true. The dramatic construction of double-take leading to final success, the sudden realization of what had gone wrong, the journey back to the deserted midnight workshop, despair transformed into triumphant vindication—these are the stuff of which melodrama is made, and Singer's melodramatic soul must have rejoiced. Indeed, it seems so much to his taste that one might be tempted to suspect him of at least a little embroidery, were it not for the fact that Zieber confirmed the account in almost every detail, the only difference being that, according to Zieber, they had actually got back to their hotel and into bed before they realized what was wrong.

"By God" (he was much addicted to profanity), "do you think so," he said. "Let's go back to the shop and try it again." Upon which we got up at one o'clock in the morning and went back to the machine, when I held my finger against the thread upon the face plate of the machine, to keep it up from the point of the needle, which caused a tight seam to be produced. We then arranged what was called the "spring pad." I mention this circumstance only to show how anxious we felt about the success of the enterprise which was to Singer especially, the "forlorn hope."

Both Singer and Zieber must have been both mentally and physically exhausted after that evening, for in addition to the lack of sleep and the tension from which they both suffered, there had been the added strain of conveying this urgency to Phelps's workmen, none of whom were themselves particularly interested in the machine, and none of whom wanted to work the long hours necessary to get it done in time. All the parts, of course, had to be handmade, and this was a long and skilled job. Jott Grant, who was one of the workmen involved, later recalled that time, with its curiously makeshift quality, and the tension in the air: "I think there were some drawings at different times, just chalked out on a piece of board or bench, or something of the kind, by Mr. Singer. When we commenced working on it, we worked night and day until eleven or twelve o'clock, until the machine was done . . . Mr. Singer was there continually, day and night, Mr. Zieber also. Mr. Phelps was there usually through the day but I don't remember of his ever being there a single night." All this work, Grant recalled, went on in secrecy behind the locked door of the room on the ground floor where Singer kept his carving machine. "One night I made the spring that drove the shuttle on the cam. It did not work just right. We experimented on it once or twice, and Singer, rather cross, ripped out to me, and I threw down my tools and told him I would not work for him any longer and left." But "I returned next morning and went to work again." Indeed, Singer's temper, always short, must have flared up more than once during those days, but he made up for these outbursts in other ways. "Mr. Singer used to perform in the shop to amuse us," Grant remembered, "theatrical performances. It was a solo: he would speak different parts at different times." In answer to the question of whether that had "materially assisted you in the construction of the sewing machine," Grant replied: "It rather amused anybody that was working night and day, especially when they were not particularly interested in the machine!" Indeed, now that they were over, Singer seems to have recalled his acting days with some nostalgia. Mrs. Phelps, who had boarded at the same house as Singer before she was married, recalled that "I had never seen him upon the stage, but occasionally he would perform a little to please the boarders and amuse the ladies. One particular time I recollect he was performing, and flourishing around with his cane he broke an astral lamp shade, which amused the family." This recollection is noted in a court record; on the blank page opposite, Singer has written rather wistfully: "*Singer a Tragedian, a good Richard, one of the best of his age.*"

But those days of what he probably liked to think of as freedom and irresponsibility were over now. He was in business, and at last there seemed a chance of success. On 7 November, Singer and Phelps optimistically composed an advertisement to be placed in newspapers. "SEWING BY MACHINERY" it proclaimed in bold letters, addressing itself to "the Journeyman Tailors, Sempstresses, Employers, and all others interested in Sewing of any description."

Several attempts have been made to produce Machines for Sewing, but they were not without many objections, and could be used to little advantage or profit.

Singer & Phelps' Belay-stitch Sewing Machine, invented by Isaac M. Singer and manufactured by Singer & Phelps, no. 19 Harvard Place, Boston, Mass., is offered to the public as a perfect machine, and will be constructed and adapted to perform any kind of work from the stitching of a fine shirt-bosom to a ship's sail, as well as some descriptions of leather. The needle is straight, and works perpendicularly upon the table of the machine, affording room and opportunity to adjust the fabric in any way—and the stitch may be regulated to any length, even to a hair's breadth.

The stitch made upon this machine is one which will not ravel or open, and the sewing can only be taken apart by cutting every second or third stitch. From 500 to 1500 stitches, according to the fabric operated upon, may be taken per minute.

To bring this machine to perfection, much labor and study has been expended upon it by the inventor, and the subscribers now offer the Belay-stitch Sewing Machine, feeling confident that it will correspond, upon trial, with the recommendation here given . . .

This machine, with ordinary care, is warranted to run one year without repairs, and will last many years. And it is so simple in its construction, and so easily regulated and managed, that any person of ordinary ability may operate it.

The price of one of these machines, which is worked by a treadle . . . is one hundred and twenty-five dollars, complete, with all appendages for operation.

These machines are so beautiful and neat in their appearance, and take up so little room, that they are an ornament to any lady's sewing apartment.

An Agent (with whom exclusive arrangements will be made) is wanted in every city and town in the United States.

On the day this advertisement appeared, the *Boston Daily Times* commented:

> Yesterday, we visited the rooms of Singer and Phelps, No. 19 Harvard Place, and personally witnessed the operation of the machine. In it, that great enemy to perfection in machinery, friction, is essentially overcome, and by the most simple means . . . This machine can be worked by any woman of common intelligence. The inventors have received orders for it to do the sewing of private families, and also from practical workmen. At the common prices for plain sewing, one of their machines will net its proprietor five or six dollars a day. Its price complete with all appendages for operation is $125. It is exceedingly neat and compact in its construction, and is in fact, the prettiest, simplest and most effective result of mechanical skill that we ever saw.

It is noticeable that the name "Jenny Lind," though mentioned in the contract, appears neither in the advertisement nor the article. In fact it was never used for any of the Singer machines. Zieber explained: "At first, I thought we should call the machine the *Jenny Lind* in honor of the famous singer whom Barnum had just brought over from Europe, but then I realized that this might drop out of fashion and I asked whether we could use his name. At first he was very unwilling to allow this, saying that he felt it dishonorable for a Shakespearean actor to concern himself with such trivialities, but in the end the play on words appealed to him, and he agreed." Obviously at this time Singer still felt that the sewing machine was a sissy thing; he, as a big man, would have preferred to be involved with something physically larger and less feminine in its associations.

However, it is apparent from the tone of the advertisement that Singer and his associates felt on top of the world. Their machine was a self-evident boon to mankind and (they thought) could not fail to be speedily recognized as such. Others, more experienced in the sewing machine business, were less sanguine. Blodgett, one of the patentees of the Lerow and Blodgett machine improved upon by Singer, and who was a tailor by trade and thus far more experienced than either Singer or Zieber in the needs and preferences of his trade, told Singer he was an idiot to try and make sewing machines to *sell.*

They would not work and the only money to be made out of the business was in selling territorial rights. Furthermore, Blodgett told Singer, he was positive that "sewing machines would never come into use. He said that he had established three factories operating with sewing-machines and that they had all failed, and that he was satisfied that the manufacture of machines would not pay." To the new partners, flushed with success, this must have sounded like the sourest of sour grapes. But if they had been capable of taking a detached view, and judging from previous experience, they would have had to admit that there was no reason to suppose that Blodgett was not absolutely right.

4

Many Inventors

Isaac Merritt Singer did not invent the sewing machine, nor did he ever claim to have done so. On the contrary, by the time he made its acquaintance in 1850, the sewing machine had been invented at least four times already. Singer's machine was the first one to work in a practical way, but that was another long and acrimonious story.

The sewing machine's history of multiple births was hardly surprising. One of the first to benefit from the introduction of labor-saving machinery at the very beginning of the Industrial Revolution had been the textile industry; Crompton's mule and the spinning jenny are perhaps the most famous of all the early machines. If spinning and weaving could be mechanized, sewing was an obvious candidate; like them it necessitated the regular repetition of the same movements. The smaller and more uniform the stitch, the better. The hand sewer's whole effort was spent in trying to complete the job as fast and neatly as possible. A hand sewer was, in effect, nothing but a human machine; the introduction of sewing machines would merely rationalize the situation.

The difficulty with inventing a device for mechanical sewing was similar to that with the mechanical writing device: to divorce the concept of the desired end from the picture of the human activity required to achieve it by hand. In both cases, the action of the machine is quite different from that of the hand performing the same task, so that a large effort of imagination was required of the early inventors.

Both the typewriter and the sewing machine, the two great

labor-saving appliances of the nineteenth century, were perfected and brought into production in the United States rather than Europe. The approach which proved in the end to be most productive suited the American rather than the European attitude to science and invention. In Europe, where such a premium was put on education and philosophy, where theoretical science had been all the rage among the elite since the polymaths of the seventeenth century, the progression from theory to practice was the natural one, and European mechanical innovators, such as Watt, Nasmyth and Brunel, were generally well versed in the theory of their particular fields. American inventors, on the other hand, tended much more to work from a situation and a need of their own: someone tinkered with the tool he was using to see if he could make it more efficient, or tried to produce a machine of his own to speed a process which was holding up the work. "Every workman seems to be continually devising some new thing to assist him in his work," reported the eminent English engineer Joseph Whitworth, on his visit to the United States in 1854. British workmen took no such attitude; there was too much unskilled labor around, and it was not in their interests to speed up jobs, any more than it was in the interests of their employers to invest in new labor-saving machinery, when labor itself was so cheap and plentiful.

The history of the sewing machine, in fact, followed the route of many other bright technical ideas of the nineteenth century (including that of interchangeable parts, which was to make possible its eventual mass production). It was initially heard of in Europe. The first patent for anything identifiable as a sewing machine was taken out in 1755 by a German mechanic living in London, Charles Weisenthal. His machine was for embroidery rather than to sew two pieces of cloth together, but he did hit on one of the essential elements in later sewing machines: the eye-pointed needle. However, he failed to recognize the vital quality of such a needle, which is that there is no need to pass it entirely through the fabric. His needle, having an eye at each end, did not have to be turned, but the device was no more labor-saving than another, patented in France in 1804, in which a common needle was used and a pair of pincers on each side of the fabric imitated the action of fingers, pulling and releasing the needle.

Neither Weisenthal nor Thomas Stone and John Henderson, patentees of the pincers machine, had been able to escape from the idea of the human hand in sewing.

The first patentee of a sewing machine resembling the type of design eventually adopted was Thomas Saint, an English cabinet-maker. In July 1790, he took out a patent on "An Entire New Method of Making and Completing Shoes, Boots, Splatterdashes, Clogs and Other Articles, by means of Tools and Machines also Invented by Me for that Purpose, and of Certain Compositions of the Nature of Japan or Varnish, which will be very advantageous in many useful Applications." Lost in the detail of this patent was a description of a machine for "stitching, quilting or sewing" which contains many features later to be incorporated in successful sewing machines: an overhanging arm with a straight, perpendicular needle; a horizontal cloth table; and thread continuously fed from a spool. The machine was to be powered by a hand crank on a shaft, activating cams which worked the various parts. There is no mention in contemporary sources of such a machine ever having been used, and its details, lost amid the rest of the patent, were not rediscovered until 1873 when a Mr. Newton Wilson, himself a manufacturer of sewing machines, came across the patent entirely by chance. "Examining some patents for boots and shoes in the library of the Patent Office," he wrote, "I came across one of ancient date, going back in fact to the last century. This was nominally for inventions of cements . . . but right in the heart of the specification was a single sheet of drawings . . . and . . . occupying the central position in the sheet, the drawing of a sewing machine . . . I gazed at that sheet with infinite interest. Here surely was the first idea of a sewing machine!" Newton Wilson actually built a machine from Saint's specifications (now in the Science Museum, London), but found that he could not make it work without some modification. It therefore seems reasonable to assume that Saint never actually built his machine.

The next incarnation of the sewing machine was in France, near the textile center of Lyons. This time the inventor was a journeyman tailor named Barthélemy Thimonnier. Although he was very poor, Thimonnier had received a certain amount of education at a Lyons *lycée*, and perhaps it was this which made him so discontented with his lot, stitching away to support a young wife and rapidly growing family in the little town of Amplepuis in the 1820s. For whatever reason, Thimonnier became obsessed with the idea that there must be some better, mechanical way to perform the monotonous task of sewing seams. Accordingly, he spent two years experimenting with one method after another, meanwhile doing barely enough tailoring to keep his family alive. And at last, he evolved a machine which did do that job: this machine was patented in France in 1830.

Thimonnier's first machine was an extraordinarily clumsy mechanism, enormously large, set in a heavy wooden frame not unlike that of a home weaver's loom, which produced a chain stitch using a barbed needle. It did not, in fact, stitch any faster than a skilled tailor did by hand; nevertheless, Thimonnier found a backer, and set up a workshop in the rue de Sèvres in Paris. By 1831, he had eighty seamstresses working there, tailoring uniforms for the French army. By now, his best machines were capable of doing a hundred stitches a minute.

Thimonnier was defeated by social circumstances entirely outside his control. The French working classes at this time were increasingly infuriated with their emperor, Louis-Philippe. They were seeing their revolution melt fruitlessly away, and the high hopes of twenty and thirty years before were being replaced with the terrifying specter of unemployment. Like so many of the working men in Britain at this time, they could see no for machinery other than to deprive them of jobs which were already scarce and underpaid. The cry of the time was, "Le machinisme—voilà l'ennemi!" Jacquard, the inventor of the famous silk loom, narrowly escaped being thrown into the Rhône at Lyons. As for Thimonnier, two hundred journeyman tailors, exasperated at the success of his workshop in the rue de Sèvres, stormed it and threw the machines out of the windows, while the inventor had to make a hurried getaway through a back door.

Thimonnier was not to be defeated so easily. He returned to Amplepuis, and spent another two years working there as a tailor while building a new and improved machine, which he called the *couso-brodeur*. Once again he found a backer, a local businessman called

Jean-Louis Magnin, and they started a company for the manufacture of sewing machines at Villefranche-sur-Saône. (The machines bore the name of Magnin, not Thimonnier.) An argument now arose which was to dog the early history of sewing machines wherever they appeared. The *couso-brodeur*, which could sew 300 stitches a minute, would, it was argued, put all the poor seamstresses and *lingères* out of work. Thimonnier rejected this argument, pointing out that the Jacquard loom had had just the opposite effect on the silk industry at Lyons. He was surprised, he said, at the amount of vilification his machine was attracting. "What is my little chain-stitch sewing machine by comparison with all these marvelous inventions, that it should be singled out for attack!" he cried bitterly. But once again he was overtaken by politics. Just as he was getting re-established in France, the troubles of 1848 threatened to break him once again, and he had to flee to England.

He was received with great honor across the Channel and was welcomed everywhere he went—a kindness which the poor man found quite overwhelming. A Manchester company even went so far as to install his machines, so that it would be true to claim that in both England and France Thimonnier's was the first machine to be used commercially. But his luck had still not changed. Thimonnier planned to exhibit his machine at the Great Exhibition staged at the Crystal Palace in London in 1851—an event attracting world-wide interest, where everything that was new in machinery, from all over the world, was to be on show. He prepared an ultra-modern version of the *couso-brodeur* especially for this occasion; but, somehow, it was broken en route from Paris to London, and lay unrepaired at his London agent's house until after the jury's examination of the machines at the Crystal Palace was over.*

* Not that sewing machines caused any great stir at the Exhibition, though several were exhibited. They seem to have been noticed only by the reporter from the Italian *Giornale di Roma*, who wrote: "A little further on, you stop before a small brass machine, about the size of a quart pot, you fancy it is a meat roaster; not at all. Ha! Ha! It is a tailor! Yes, a veritable stitcher. Present a piece of cloth to it; suddenly it becomes agitated, it twists about, screams audibly—a pair of scissors are projected forth—the cloth is cut; a needle sets to work; and lo and behold, the process of sewing goes on with feverish

This final straw really broke poor Thimonnier's back. He could never quite persuade himself that the accident was not part of some sinister plot engineered by his competitors to destroy him. He gave up, returning to Amplepuis, ill and worn out, with 3.25 francs in his pocket, and died a broken man in August 1857.

In 1834, while Thimonnier was licking his wounds in Amplepuis after his first defeat at the hands of the tailors, the sewing machine was invented for the third time, in Amos Street, New York. The maker of this machine was Walter Hunt, a mechanic and inventor of real genius. Among Hunt's other inventions were an improved oil lamp, an ice-plow, the first rotary street-sweeping machine, the first home knife sharpener, the safety pin (to which he immediately sold the rights for $400 cash in order to pay off a $15 debt) and the breech-loading rifle. Here was a man with real mechanical imagination, and it is no coincidence that he was the first designer to get away totally from the idea of hand sewing and produce a sewing method designed specifically for the machine.

Until this time all the machines designed had produced a chain stitch, using one thread and catching each loop in the one following it. The disadvantage of this stitch was that it was easily unraveled; if a stitch tore, you only needed to pull the thread for the whole seam to come apart. Hunt devised the stitch known as the lockstitch which is still used in modern sewing machines. This involves using two spools of thread, one above the fabric, one below, with a shuttle to push the lower thread through the loop made by the upper one as it is pushed through the fabric by an eye-pointed needle. The needle with the upper thread then retracts, and the shuttle returns to await the next stitch. This was the stitch subsequently used in all success-

---

activity, and before you have taken three steps a pair of inexpressibles are thrown to your feet, and the impatient machine, all fretting and fuming, seems to expect a second piece of cloth at your hands. Take care, however, as you pass along, that this most industrious of all possible machines does not lay hold of your coat or greatcoat; if it touches even the hem of the garment it is enough—it is appropriated, the scissors are whipped out, and with its accustomed intelligence the machine sets to work, and in a twinkling another pair is produced of that article of attire, for which the English have as yet been unable to discover a name in their most comprehensive vocabulary."

ful sewing machines, including that patented by Lerow and Blodgett which Singer saw in Orson Phelps's workshop.

Hunt's invention of this stitch was to prove of some importance in the patent battles of the 1850s: but at first he was not particularly interested in commercializing or even patenting his machine. He sold a half-interest in it to George Arrowsmith, a blacksmith with a workshop in Gold Street who was the employer of Hunt's brother Adoniram. In 1835, Adoniram was sent on business to Baltimore and, taking Hunt's sewing machine with him, he demonstrated it at the house of a friend, Joel Johnson, to whom he wrote in 1836: "I made that sewing machine that I had at your house work like a charm," and added that he wanted to build a stronger machine of iron (rather than wood). In 1838, Hunt suggested that his daughter Caroline, then aged fifteen, might like to go into business manufacturing corsets with the aid of his sewing machine, since it would take the effort out of the heavy stitching needed. Caroline, however, turned up her nose at this suggestion, since she was, like her father, a good Quaker, and thought the machine would do a great deal of harm to the thousands of poor seamstresses who made such a precarious living. Hunt accepted this argument, and took no further interest in his sewing machine.

The very next year, the ground was prepared for yet another genesis of this phoenix among machines. The scene this time was a machine shop in Boston—belonging not to Orson Phelps but to one Ari Davis. Among the young apprentices there was Elias Howe, at this time twenty years old, the son of a poor New England farmer and miller. Howe suffered from congenital lameness, which made the life of a farmworker difficult for him. After spending some years in his father's mills, he set off for Cambridge, where he took casual work in a number of machine shops, ending up in that of Ari Davis.

In 1839, two men, an inventor and his financial backer, came to Davis with plans for a knitting machine. The inventor was having difficulty with his design, and thought that perhaps Davis could help him over the difficulties and get his machine to work.*

* He was rather late off the mark, since the first knitting machine was invented in 1598 by William Lee, who "acquired an aversion to hand-knitting because a young woman to whom he was paying his addresses seemed, when

The ensuing conversation between the inventor and Davis was recounted by Howe to James Parton some twenty-eight years after the event:

The shop, resolving itself into a committee of the whole, gathered about the knitting machine and its proprietor, and were listening to an explanation of its principle, when Davis, in his wild, extravagant way, broke in with these words: "What are you bothering yourselves with a knitting machine for? Why don't you make a sewing machine?"

"I wish I could," said the capitalist; "But it can't be done."

"O yes, it can," said Davis; "I can make a sewing machine myself."

"Well," said the other, "you do it, Davis, and I'll insure you an independent fortune."

Davis, despite his boast, never did make a sewing machine, and we have no record as to whether he made a fortune. But this conversation deeply affected the young Elias Howe, who was much impressed with the imposing figure of the capitalist. During the next few years,

---

he visited her, to be always more mindful of her knitting than of his presence. He therefore invented the Stocking Frame, which Queen Elizabeth came to see in action. But she was disappointed by the coarseness of the work, having hoped it would make silk stockings, and refused to grant the monopoly he asked for. Lee altered the machine and produced a pair of silk stockings in 1598, which he presented to the Queen, but Elizabeth now feared that the invention would prejudice hand-knitters and it was consequently discouraged. As King James took up a similar attitude, Lee went to France, where he had been promised great rewards by King Henry IV, but the latter's assassination disappointed these hopes, and Lee died of grief in Paris, in 1610." John Aubrey, in his *Brief Lives*, did a Life of Lee, and shows that the knitting machine had to face just the same discouragement from those it was meant to benefit as did the sewing machine later: "It ought never to be forgot, what our ingenious Country-Man Sir Christopher Wren proposed to the Silk-Stocking Weavers of London, viz. a way to weave seven pair or nine pair of stockings at once (it must be an odd number). He demanded £400 for his Invention; but the weavers refused it, because they were poor: and besides, they said, it would spoil their Trade; perhaps they did not consider the Proverb, that Light Gaines, with quick returns, make heavy Purses. Sir Christopher was so noble, seeing they would not adventure so much money, he breaks the Model of the Engine all to pieces, before their faces."

he spent a great deal of his spare time pondering the problems of a sewing machine. Quite soon his thoughts acquired a new urgency, because in 1840 he married, and soon children began to arrive. Howe was still working as a journeyman machinist, and earning nine dollars a week. This was hardly enough, even in those days, to feed, clothe and house a wife and three children: he would go to bed at night wishing only, as he later reminisced, "to lie in bed for ever and ever." The idea of an independent fortune began to take on new and concrete attractions.

Howe began to work seriously on the idea of the sewing machine in 1843. It took him some time to escape from the old trap of copying the action of a hand sewer: his first device was a needle pointed at both ends with an eye in the middle. But one day in 1844, he realized that the machine might have capacities which the hand did not, and the idea came to him of using two threads and forming a stitch by way of a shuttle—just as Hunt had done ten years earlier. By October 1844 he had constructed a rough model of wood and wire which was sufficient to convince him that his idea would work. However, he now struck the perennial snag of the poor inventor. How was he to obtain the time and money he needed to construct a finished working model? Parton, writing in 1867, reckoned that "at the present time, with a machine before him for a model, a good mechanic could not, with his ordinary tools, construct a sewing machine in less than two months, nor at a less expense than three hundred dollars." Howe, who had his model only in his head, could certainly not expect to take any less time with his machine, and there were his wife and children to support into the bargain.

At this time Howe was no longer working as a mechanic. His father had moved to Cambridge where he was trying to establish an invention of his brother's, a machine for cutting palm leaf into strips for use in making hats. Elias and his father were sharing a house, and Elias had moved some tools and a lathe into the garret, where he spent what time he could on his own work. The family circumstances were further reduced when the leaf-cutting machine was destroyed in a fire, and it was felt that Elias could be doing more for his family than fiddling around in the garret. However, he now had a stroke of luck—an unusual event in Elias Howe's life. An old school friend named George Fisher was living in Cambridge. Fisher had just inherited some property and was willing to invest some of it in the possibilities of Howe's new machine. Late in 1844, therefore, the two went into partnership. The terms of the partnership were that Fisher agreed to board Howe and his family in his house and let Howe use the garret for a workshop. He would also supply $500 for materials and tools. In return, he was to become proprietor of one half the patent, should the machine prove to be worth patenting. In December 1844, the Howes moved into Fisher's house and Elias began to work. Nobody else in Cambridge shared Fisher's confidence in Howe and his machine. In one of the later patent suits Fisher testified: "I believe I was the only one of his neighbors and friends in Cambridge that had any confidence in the success of the invention. He was generally looked upon as very visionary in undertaking anything of the kind, and I was thought very foolish in assisting him."

Throughout the winter of 1844–45 Howe worked at his machine. "His conception of what he intended to produce was so clear and complete," says Parton, his biographer, "that he . . . worked on with almost as much certainty and steadiness as though he had a model before him. In April he sewed a seam by his machine. By the middle of May 1845, he had completed his work. In July, he sewed by his machine all the seams of two suits of woolen clothes, one suit for Mr. Fisher and the other for himself, the sewing of both of which outlasted the cloth."

Howe now faced the most difficult task of all, which was to persuade the public, and above all the tailors, first that his machine would really perform as well as he claimed it would, and second that it was worth buying. One thing was certain: nobody was going to believe anything just because he said it was true. Accordingly, he set himself and his machine up as a public show in the Quincy Hall Clothing Manufactory, Boston, where he offered to sew up any seam which might be brought to him on his little machine. The machine at this time sewed 250 stitches a minute, which was about seven times as fast as hand sewing. Howe cheerfully sewed any type of seam, even the clumsiest—such as sewing the thick, heavy skirts of frock coats to the bodices. Having successfully accomplished anything which might be presented to him in this line, he challenged five of the fastest seamstresses who could be found to a race with his machine. He would finish all five of his seams before any of them could finish one. The seamstresses agreed, and sewed their seams as fast as they could—"much faster than they were in the habit of sewing." Nevertheless, Howe finished his seams first and a tailor umpire judged that "the work done on the machine was the neatest and strongest."

Howe must now have felt that he could, and indeed need, do no more. What more could be asked of any machine? He and Fisher sat back and confidently waited for the orders to roll in. There were none. The usual reasons were adduced: journeyman tailors would object; seamstresses would be put out of work and reduced to poverty; tailors were doing nicely as they were, thank you, and saw no need for change. The machine was all very well, but it couldn't do all the jobs a tailor had to do, since tailoring was more than a matter of sewing seams. Above all, it would be hard to justify the expenditure. A tailor of any size would have to buy thirty or forty machines to equip his workshop, and Howe would be unable to provide them for less than $300 apiece. This last was the greatest objection, and as far as Howe could see, there was no way round it.

Howe therefore resigned himself to the fact that there was apparently no prospect of making any money out of his machine. The capitalist of 1839 had been wrong: no independent fortune, not even a living, was in the offing. So he took a job as an "engineer"—in other words, a driver—with one of the railroads terminating in Boston, and in his spare time worked in the garret upon a second model for the Patent Office in Washington. When this and the documents were ready, in the summer of 1846, he and Fisher made the journey to Washington, where the patent was issued on 10 September. The machine once again provided amusement for the Washington crowds, and was noticed in the 26 September issue of *Scientific American* on its "New Inventions" page:

We have heretofore noticed the extraordinary invention by Mr. Elias Howe, Jr., of Cambridge, Mass.—a machine that sews beautiful and strong seams in cloth as rapid as nine tailors. We are not yet prepared to furnish a full description of the machine, but the following claims, in the words of the patentee, may give some idea of the various parts in combination . . .

"I claim the lifting of the thread that passes through the needle eye by the lifting rod, for the purpose of forming a loop of loose thread that is to be subsequently drawn in by the passage of the shuttle; said lifting rod being furnished with a lifting pin, and governed in its motions by the guide pieces and other devices.

"I claim the holding of the thread that is given out by the shuttle, so as to prevent its unwinding from the shuttle bobbin, after the shuttle has passed through the loop, said thread being held by means of the lever, or clipping piece.

"I claim the manner of arranging and combining the small lever, with the sliding box in combination with the spring piece, for the purpose of tightening the stitch as the needle is retracted.

"I claim the holding of the cloth to be sewn, by the use of a baster plate, furnished with points for that purpose, and with holes enabling it to operate as a rack, thereby carrying the cloth forward, and dispensing altogether with the necessity of basting the parts together."

Howe's model is acknowledged to be one of the most beautiful ever presented to the Patent Office. But once again, neither its beauty nor its efficacy showed any signs of so much as recovering its cost. The two young men returned to Boston. George Fisher was by now resigned to having invested $2,000 in a dead loss, and Howe returned to his father's house. Not unnaturally, he was unwilling to let his brainchild die without a fight. Since he owned nothing at all, he had nothing to risk and everything to gain in trying to promote his machine. He resolved to try his luck in England, and see if the sewing machine might find a more profitable reception on the other side of the Atlantic. Accordingly, in October 1846, his brother, Amasa B. Howe, with some financial help from their father, took a steerage passage in a sailing packet to London, carrying a machine with him. He seemed to strike it lucky at once. A Mr. William Thomas, a corset manufacturer with big premises in Cheapside, London, was interested in the machine. This seemed a hopeful prospect, since Thomas employed at that time, or so he said, five thousand people in the manufacture of corsets, umbrellas, valises, carpet bags and shoes. Amasa Howe sold him the machine he had with him for £250, which also bought the right to use as many others as he wished in his own business. There was also a verbal agreement that William Thomas was to patent the machine in England—which he did—and if the machine caught on, he was to pay the inventor a royalty of £3 on every machine sold. This he declined to do, though he did claim a royalty of £10 himself on each machine sold. He was also extremely vigilant about suing anyone he considered to be infringing his patent. The result of this was to delay the widespread introduction of the sewing machine into England until Thomas's patent expired at the end of fourteen years. Elias Howe later reckoned that Thomas's £250 investment must have yielded him more than a million dollars, and although his calculations may have been exaggerated by his understandable bitterness, there can be no doubt that Thomas drove a very hard bargain with poor Howe. Like the capitalist, he recognized the prospect of a fortune when he saw it. Unlike Elias Howe or his brother, he also knew how to make the most of the situation.

Howe's machine was not yet ideal for corset-making however, so Amasa returned to Cambridge armed not only with the £250 (which was eaten up almost at once; the Howes had been living in anticipation of it for some time) but with an offer to pay Elias the princely sum of £3 a week if he would go over to England to make the necessary alterations. Since there was no work in prospect in Cambridge, Elias agreed, and he and Amasa once more took ship for England, traveling in the steerage and catering for themselves en route as was customary for steerage passengers at that time. Elias's wife and three children followed shortly after.

The work of adapting the machine to corsetry took eight months, after which time Howe's usefulness to Thomas was exhausted and he threatened to become an embarrassment. Thomas thereupon set about making life impossible for the inventor, getting him to act as a general repairman around the workshop and treating him with great insolence. Howe naturally resented this, and Thomas thereupon sacked him. Howe now found himself in an even worse situation than before: almost penniless once again and this time in a strange country.

He moved his family to cheaper rooms, and began to build yet another machine in order to raise some money. He decided to send his family home while he still could, and follow them when he had built and sold the machine. His state at this time was described by one of his London friends, a coachmaker called Charles Inglis, in testimony during one of the patent suits:

> Before his wife left London he had frequently borrowed money from me in sums of five pounds, and requested me to get him credit for provisions. On the evening of Mrs. Howe's departure, the night was very wet and stormy, and, her health being delicate, she was unable to walk to the ship. He had no money to pay the cab-hire, and he borrowed a few shillings from me to pay it, which he re-paid by pledging some of his clothing. Some linen came home from his washerwoman for his wife and children on the day of her departure. She could not take it on account of not having money to pay the woman.

Three or four months later Howe finished his machine, which was worth £50. The best offer he could get was for £5 from a poor work-man who needed time to pay. In order to pay his debts and his passage home, Howe pawned his first machine and his letters patent. Once again, he crossed the Atlantic in the steerage hold.

He landed in New York in April 1849, with half a crown in his pocket to show for four years of unremitting hard work. He im-mediately took a job in a machine shop, where news reached him that his wife was dying of consumption in Cambridge. He had not even the money to travel to her, but a remittance of ten dollars from his father got him to the bedside in time to see her before she died. He had to borrow a suit from his brother-in-law for the funeral. Friends noticed that he seemed "extremely downcast and worn"—and no wonder. The next news he received was that the ship containing all his house-hold goods had been wrecked off Cape Cod, and everything was lost. Howe worked no more upon his sewing machine, but took a job in a machine shop for a weekly wage, which his family, and probably George Fisher, considered the most sensible thing he had done for years.

5

The Sewing Machine War

Going by previous experience, then, the logical thing at this time must have been to assume that the tailor Blodgett was right and Howe's capitalist quite wrong in their assessments of sewing machine potential. But was that logic? For logic, in another sense, pointed in quite the opposite direction. Looked at in the abstract, in terms purely of ideas and markets, the sewing machine *could not* fail. Anyone who wears clothes must sew or employ people to sew for him; and anyone who sews must surely welcome a machine which does the boring, laborious operations faster, better and more easily than they can be done by hand. The savings in time and effort were potentially quite enormous, as was shown by figures prepared by the Wheeler and Wilson Company ten years later. They compared the time taken to complete various garments by hand and using a machine. Gentlemen's shirts, they reckoned, took 14 hours 26 minutes to stitch by hand; 1 hour and 16 minutes using a machine. A frock coat which took $16\frac{1}{2}$ hours by hand was completed in a mere $2\frac{1}{2}$ using a machine. A merino dress taking $8\frac{1}{4}$ hours by hand took just over 1 hour on the machine. The comparisons are almost absurd.

Surely the market had to be there? And what a market! There were in 1850 an estimated five thousand shirt sewers in New York City alone. Their lot was a miserable one and was generally known to be such. "We know of no class of workwomen," reported the *New York Herald* in 1853, "who are more poorly paid for their work or who suffer more privation and hardship." About a quarter of these women made a maximum of $1.50 a week. Rather more could make $2.00; and the average wage was about $2.50 a week. Not more than one fifth of the shirtwomen earned $4.00, $5.00 or at the very most $6.00 a week. Two and a half dollars, remarked the *Herald*, was scarcely adequate to support one person for a week. "A tailor," it went on, "gets five dollars for a coat taking two days, a shirtwoman gets a maximum of one and a half dollars, working twelve or fourteen hours a day." These were the women of whom Thomas Hood wrote his popular—and essentially factual—ditty, the "Song of the Shirt":

> With fingers weary and worn,
> With eyelids heavy and red,
> A woman sat, in unwomanly rags,
> Plying her needle and thread,
> Stitch! Stitch! Stitch!
> In poverty, hunger and dirt;
> And still with a voice of dolorous pitch—
> Would that its tone could reach the rich!—
> She sang this "Song of the Shirt!"

Despite the frightful conditions and prospects, the supply of sewing women, on both sides of the Atlantic, was inexhaustible. For in the mid-nineteenth century there were very few ways by which a woman, left for one reason or another to fend for herself, could earn a living. She could teach, but that usually required a certain amount of capital with which to start a school. In Europe, she could become a governess or a lady's companion; but the demand for governesses was low in the United States, and the subservience and dependence of the job, which was hard enough for English girls to accept—as many novels about governesses made only too clear—was absolute anathema to independent Americans. She could start a shop, but again, that required capital. She could go on the streets—a common fate for poor girls. Or she could take in sewing, and attempt to keep up a pitiable front of respectability in the face of impossible conditions. Because of the lack of choice of employment and the desperate necessity to keep up a front of respectability, employers could—and did—grind the poor sewing women down with little fear of any protest or reaction.

The case of one such girl is specifically cited by Karl Marx as an example of acute industrial hardship:

In the last week of June, 1863, all the London daily papers published a paragraph with the "sensational" heading, "Death from simple over-work!" It dealt with the death of the milliner, Mary Anne Walkley, 20 years of age, employed in a highly respectable dressmaking establishment, exploited by a lady with the pleasant name of Elise . . . This girl worked, on an average, 16½ hours, during the season often 30 hours, without break, whilst her failing labour-power was revived by occasional supplies of sherry, port or coffee. It was just now the height of the season. It was necessary to conjure up in the twinkling of an eye the gorgeous dresses for the noble ladies bidden to the ball in honour of the newly-imported Princess of Wales. Mary Anne Walkley had worked without intermission for 26½ hours, with sixty other girls, thirty in one room, that only afforded one-third of the cubic feet of air required for them. At night, they slept in pairs in one of the stifling holes into which the bedroom was divided by partitions of board . . . Mary Anne Walkley fell ill on Friday, died on Sunday, without, to the astonishment of Madam Elise, having previously completed the work in hand. The doctor . . . called too late to the death-bed, duly bore witness before the coroner's jury that "Mary Anne Walkley had died from long hours of work in an over-crowded work-room, and a too-small and badly-ventilated bedroom." "Our white slaves," cried the *Morning Star*, the organ of the Free-traders, Cobden and Bright, "our white slaves, who are toiled into the grave, for the most part silently pine and die."

For such women as these, the obvious effect of the sewing machine—if they could only afford to buy one—might be expected to be wholly beneficial (as it indeed was when they eventually came around to it). They could complete the same work in a fraction of the time. If the work available did not actually increase, they could still expect to earn as much as ever, without having to spend all their waking hours crouched over a needle. But this was not how they saw it. As late as 1858, the spirit of the Luddites and of the enraged tailors who destroyed poor Thimonnier's machines still held sway in the Shirt Sewers' and Seamstresses' Union. An address to that union in April 1858 pointed out that there were more than forty thousand sewing women in New York (a much larger figure than that quoted by the *Herald* five years earlier, but then perhaps not all these were specialized shirt sewers) and of that number, probably not more than three thousand found employment during the cold winter of 1857-58. "They might well imagine the disastrous consequences of such a result to a large portion of the community," continued the address, "and they might also be aware that the sewing machines, out of which those houses which manufactured them were reaping such large fortunes, were now diffused to such an extent that ere long it was not improbable that that peculiar branch of industry which exclusively belonged to women—that industry which developed itself in the facile and pliant use of the fingers—would be totally extinguished."

The one clear train of thought in this rather muddled tirade is easy to discern: women do not use machines, and will only get work for which no machinery is available. A very large part of Singer's subsequent marketing campaigns was (as we shall see) specifically directed toward combating this attitude, which was deeply ingrained in the social life and aspirations of the time. What this meant was that, in the early 1850s, the main potential market for the sewing machine neither wanted it nor could afford it. Moreover, such attitudes were bolstered rather than undermined by most of the sewing machines currently on offer. It wasn't just an abstract sewing machine that potential customers didn't want; they had seen what was around, and they definitely didn't want *that*. When Singer, Phelps and Zieber entered the sewing machine market in 1850, they had to batter against a wall of sales resistance which had been built up as a result of a number of sewing machines on the market which were simply more trouble than they were worth. "I met with continual objections to the introduction of my machine," wrote Singer of his early sales trips, "from persons who had bought those of prior inventors and had thrown them aside as useless, and in some cases was showed out of the stores where I called as soon as my business was made known by me."

For the sewing machine had not died with Elias Howe's interest in it. When Howe returned to America, he was surprised to discover that, in the years since he sailed for England, sewing machines had become quite famous in America. Mechanics had read of his device or seen it demonstrated, and had turned their hand to producing something similar. The Lerow and Blodgett machine which had been the basis for Singer's improvements was one such piece of work. Sewing machines were demonstrated at the Massachusetts Charitable Mechanics Exhibition in 1850: indeed, a Lerow and Blodgett machine won a silver medal there and another, by A. B. Wilson, won a bronze medal. (Wilson's machine, in fact, contained some unique and excellent features, and he went on to co-found the very successful Wheeler and Wilson Sewing Machine Company.)

Howe's attention was drawn to these machines by a poster which he noticed at Ithaca, New York, in May 1849, a few weeks after his return from Europe:

A GREAT
CURIOSITY
The
YANKEE SEWING-MACHINE
is now
EXHIBITING
AT THIS PLACE
FROM
8 a.m. to 5 p.m.

Howe made it his business to inspect some of these machines which were no longer just a curiosity but were actually coming into use in a few tailors' shops. They were not popular because (like the ones made and repaired by Orson Phelps) they were very expensive and always breaking down. Howe considered them much inferior to his own machine. Nevertheless, they were an infringement of his patent, and to be cheated in the United States as well as in England was more than the poor man could stand. If there was any money, any money at all, to be made out of sewing machines, he was determined to have a share of it. As things stood, however, he was at a disadvantage, because his letters patent were still with the pawnbroker on the other side of the Atlantic. What was more, a hundred dollars was needed for their redemption, and he had reached the end of his friends' patience, not to say their pockets, on the subject of his sewing machine. But Howe was determined. Somehow the hundred dollars was raised, and the Hon. Anson Burlingame, who was going to London, was prevailed upon to hunt up the pawnbroker in the wilds of Surrey where Howe had been living. He was successful, and sent the papers home in the autumn.

Elias Howe was now all set to take up what was to be his main preoccupation—indeed, his main occupation—for the next several years: namely, suing the infringers of his patent for royalties. One observer remarked, rather unfairly, that "he litigated his way to fortune and fame"—but he would certainly never have made his fortune without litigation. However, litigation costs money, and Howe now needed yet more financial backing if he was to take up his claims. He approached George Fisher, who had helped him so much in the past and who held a half-share in the patent, but, not surprisingly, Fisher was not prepared to sink any more time or money into Howe's machine. He was willing to sell his half of the patent for what it had cost him, and nothing more would he do.

Luckily for Howe, someone still existed who was prepared to gamble upon the possibility of founding a fortune on the sewing machine, even when confronted with its history to date. This was George W. Bliss, who agreed to buy Fisher's half of the patent and advance the money for the lawsuits, but only against the security of a mortgage on Howe's father's farm. With exemplary faith or resignation the old man agreed, and Elias Howe took up his cudgels.

Meanwhile, unaware for the moment that nemesis was poised to strike, Singer, Phelps and Zieber set about marketing their machine. It had one great advantage over all previous machines: it worked successfully in practice. Singer had overcome the main defect of Howe's machine, which was that the cloth was fed through by means of a "baster plate" of limited size to which it was attached. When the machine reached the end of the plate, it had to be detached and brought back, so that it was impossible to sew a long continuous seam, to sew curved seams or turn corners. Singer's machine permitted all these things. There are ten essential features to a practical working sewing machine—the lockstitch, an eye-pointed needle, a shuttle for the second thread (vibratory or double-pointed), continuous thread from spools, a horizontal table, an overhanging arm, continuous feed (synchronous with the needle motions), thread or tension controls that give slack thread as needed, a presser foot, and the ability to sew in a straight or curving line—and only Singer's machine incorporated them all, although he was the inventor of only the last two. But how was he to make people believe him when so many had cried "Wolf!" already?

The years of touring the country as a traveling actor had perhaps not been wasted after all. Singer was no retiring and unsociable mechanic. Where Phineas T. Barnum had shown the way, there Singer could follow. Where Barnum had exhibited his freaks, there Singer could exhibit his machine, hoping to score as great a success with his Jenny Lind as Barnum, even then, was having with *his*. Singer and his machine barnstormed the country, giving free on-the-spot demonstrations wherever he could, in hired halls, at fairs and carnivals. As a theatrical extra, he gave heartrending recitations of the "Song of the Shirt." In addition to this, he advertised, sent out agents, inserted articles in newspapers, and generally made as much noise as he possibly could. On the road, he evolved yet another innovation: a packing case for the machine which could also double as a table and treadle base. He never patented this idea—presumably it never occurred to him to do so, since it was not part of the intrinsic mechanism, and only realized his omission years later, when the idea was in general use and one of his competitors triumphantly reminded him of its origin. No doubt he cursed and swore, as he habitually did—it was a characteristic noted by all his acquaintances—and if ever there was an occasion for it, that was it.

Back in Boston, things were not proceeding calmly. The first difficulty occurred right after the new machine had been so triumphantly constructed. Singer and Phelps had several violent arguments as to who should apply for the new patent—whether it should be both of them or Singer alone. Singer was stubborn and absolutely determined that the patent should be his, and Zieber, "for the sake of going along quietly" (as he later said), advised Phelps to submit. As for Phelps, he seems to have found himself completely out of his depth in the whole affair, both then and later. He was, said Zieber, "a good honest kindhearted man incapable of doing any one harm, and he used to say to me, with tears in his eyes, 'Mr. Zieber, I really do not know what to make of Singer. I never saw a man like him. I am quite disheartened at the way he is going on.' " Unfortunately, a gentle disposition tends to be a disadvantage in business. Phelps seems to have spent most of the next two years in a state of bewilderment, always vaguely aware that he was not getting his due but unable to figure out how the situation might be changed.

Having won this argument, Singer immediately spirited away the new machine to New York, explaining to Phelps that he wanted to show it to his friends, and to have the specifications and drawings made out to obtain a patent. He arrived home, as his son Augustus remembered, on about the first of October 1850. The date was fixed in Gus's memory by the fact that Mary Ann was once again due to give birth; in fact she had her baby on 6 October, a boy, Charles, who was to live only four years. Gus remembered that "at the time the physician and nurse were attending my mother, this sewing machine was in the same room, that is, my father's room, and the doctor noticed it, and it was covered up. I saw it operated upon, and experimented upon, and sewed upon, in my mother's room." Gus at the time was enrolled in a school close by but, as the only servant had just left, he was kept back from school in order to look after his brothers and sisters and generally help out at home. He was now thirteen, and it turned out that this was to be the last regular schooling he would get for some years.

Singer at this point was still trying to find a satisfactory means of keeping the thread out of the way as the needle descended, and he asked Gus for something that would do as a spring. "I hadn't any idea of what he wanted, but I said I had a gun with a spring in it, and I went and got a small toy gun that belonged to a brother younger than myself, which was out of order—in fact I took it apart myself to see what was inside of it—and I gave him the special spring that was placed in the bottom of the gun to force out the arrow." The sewing machine, says Gus, "seemed to give him more satisfaction after I gave him the spring, and he invited several persons to come in and see the machine, and several did come. The machine was put in the box it came here in, and one of the expressmen called and took it away."

Zieber had advanced the very last of his money in order to pay for the costs of the patent and the manufacture of the first two or three machines. The first of these was brought to the Singer apartment in New York some weeks later. The machine was already spoken for. Singer, while buying some clothes for his sons at Smith and Conant's clothing shop on Broadway, "took occasion to speak of the sewing machine I had invented. After some conversation they agreed to take two machines from me at $125 each." Gus was deputed to go out and find a spring cart to deliver the first of these.

> I could not get such a one as I wanted, and I got one that was without springs. I remember that it was a holiday—that it was Thanksgiving Day—the last Thursday in November. I remember the stores were closed; it was towards night, a windy and cold day, and my father reprimanded me for not getting a cart with springs, because the jolting of the cart was detrimental to the fine machinery of the sewing machine. That machine afterwards—I don't know how many days, but a very few days afterwards—was put upon a table in the store belonging to Smith and Conant, and another machine arrived about that time, or at that time, and was put up alongside of it, both at the same window. A young lady arrived from Boston, a tailoress by trade, whose name was King. She operated upon one machine and I upon the other. She did the sewing about the clothing that was complicated—she understood it—I did the plain, simple sewing.

Gus remained in the store window operating his machine until after New Year's Day, 1851. He particularly remembered the occasion because he got no present or reward for all his help—something, as he said, "that would be very likely to impress itself upon a boy's memory."

If Gus got no present it was because Singer was still desperately short of money, as were Phelps and Zieber. This temporary shortage was to cost them very dearly in the long run, for it was at this time that Elias Howe first got wind of the machines manufactured by I. M. Singer & Co., as the firm was called. Howe actually saw the machines being worked by Gus and Miss King in the window of Smith and Conant's. The sight did not please him and he called in to complain that Singer's machine was an infringement of his patent. Negotiations were begun in Boston between the two sides. Singer and Zieber were by no means unaware of the potential importance of such a deal: on the contrary, they thought that the exclusive right to use Howe's patent would ensure them a monopoly in the sewing machine business. However, Howe was asking $2,000 for such rights, and this was money they simply did not have; indeed, they had no money at all. In the end, Singer lost his temper, possibly from frustration at seeing such a golden opportunity slipping away, quarreled with Howe and threatened to kick him down the steps of the machine shop. "Mr. Howe," said Zieber, "lived to be thankful for the exhibition of Singer's amiable disposition on that occasion."

Meanwhile it was hard for the partners to raise money to exist from day to day, let alone buy out anyone else's rights. Before profits could come rolling in, materials had to be bought (although some of this could be managed on credit) and workmen had to be paid. Zieber managed to raise an additional $620 from various friends and from his brother in Philadelphia. Singer also managed to raise some money from one of his brothers—the same brother John who had bought a plot of land near Adam Singer's house for $50 in 1818 and sold it for $40 the next year. John had become a sea captain, "a rough, muscular, illiterate fellow"—there was obviously some family resemblance— who fetched up with his stewardess, Joanna Shaw, "a rather slattern woman," on a deserted island near Brazos Santiago harbor, Brownsville, in southwest Texas. There he became "as well known as any person on the frontier, where he was everywhere known as the brother of the sewing-machine man." He became friendly with the Brownsville postmaster, and one day told him this story:

It is not forty years since I, then a youth, went to sea. I left at home a boy brother whom I have seen but once since. It was several years afterwards when I met him, then a young man, on the streets of New York City. He told me that one Elias Howe, of

Connecticut, had evolved an idea of a sewing machine, but it was an idea only—not a practical machine, for use in families. He, himself, had invented a machine which could do the work, and which would become one of the greatest mechanical improvements of the age, but he had not the means to secure the patent and put it on the market. He added that with $500 he could soon make his fortune.

I shared his faith—he shared my purse. I had $500 and gave it to him. Since that time I have not seen him or my money. I have heard the results and am satisfied. Now [handing the postmaster a letter], read that letter from my brother and answer it in my name, telling him I shall want the money to build a steamship to run between Brazos and New Orleans in opposition to Harris and Morgan's ships.

The postmaster continued: "The letter was indeed from his brother, reminding him of the $500 loaned, and the realization of more than he had anticipated. It closed by saying, 'though you probably possess a competence, you may desire to adventure to some enterprise and lack the means. If so, draw on me for $150,000 and do what you desire.' I wrote the letter for him, but, the war coming on, the investment was not made."

Such munificent returns, however, lay far in the future. The present, even with the help of brothers and friends, looked bleak. Not only the workmen but also Singer's ever-larger family clamored for money. The partners, in desperation, advertised in the *Boston Daily Times* for a partner with $1,000; they would have sold a one-fourth interest for that amount of cash to be put into the business. There were, however, no takers.

Once again, it looked as if Singer had been chasing mirages. "Singer was very much discouraged at this time," wrote Zieber, "and so distressed for want of money that he offered to sell out all his interest in the business for $1,500." Zieber persuaded him not to take this drastic step. They kept going on what they could raise haphazardly.

Phelps, meanwhile, went on making machines as fast as he could. Several had been sold, ten or twenty were under construction, and

stock had been acquired to make more at the Boston shop, when
Singer, returning from New York a few days before Christmas, 1850,
informed Phelps that they must cease operations. "I asked him why,"
recalled Phelps.

> He said they had got no money. At that time we had expended
> about $400, as near as I can judge. I asked him why the thing had
> been represented to me in that way—why I had been told that
> Mr. Zieber had $80,000. His answer was that Mr. Zieber had
> property, but could not turn it to money, and we must con-
> sequently stop. I told him I thought it was a little singular that
> the thing should come to so sudden a termination, but at any rate,
> the business must go on at all hazards—it would not do to stop it.

Singer now determined to maneuver Phelps out of his one-third
partnership. The reason he gave was that "Phelps, who was an intemperate, was a great clog upon the business." The discussions
about closing up and selling out which he had just been having with
Phelps and Zieber had, in fact, represented the low point of the
company's early operations; just after this, orders began at last to
come in. The first of these was to supply a firm of shirtmakers with
thirty machines at $100 each, a contract worth $3,000. Singer went
up to New Haven to demonstrate a machine and conclude the deal,
on the strength of which he received $1,200 in advance. He gave
$200 of this to Zieber and kept the rest for himself, "which Phelps
and I thought very strange," commented Zieber, "though we said
nothing about it. Perhaps the possession of so much money, which he
had not been used to, was a great pleasure to him, and we did not
care to deprive him of the enjoyment, so long as it was not wanted in
the manufactory."

The purpose of keeping back the money, however, was soon
revealed. After a short time, Singer proposed to Zieber that they
should buy Phelps out with it. Zieber declined to do so, "because, as
I told him, we wanted all the money we had to remain in the business.
Every dollar we could get was worth ten dollars to us, then." Singer,
however, was determined, and raised the matter again with Zieber.
"I am determined to have Phelps out," he said. "By God, he shall
remain in the business no longer." "He took every occasion to

quarrel with Phelps," wrote Zieber, "and behaved to him in the most brutal and insulting manner." Once again Zieber agreed to go along with Singer in an action of which he later said he disapproved. If we are to believe him, he did it purely for the sake of peace, but it is hard to credit that he had no thoughts for his own possible profit in the affair. It is probably fair to say he would never have taken such a course of action on his own initiative. However, Singer, not Zieber, was the one with initiative, and they proposed to Phelps that they should buy out his share for $1,000 down and another $3,000 to be paid in installments. In return, they promised that if the business was successful Phelps's shop would be kept busy. There was also a verbal understanding (and it is typical of Phelps's haphazard approach to business that it was not a written one) that the shop would be kept in operation at least for the duration of the partnership, which was for the next five years, "and that I should probably have constant employment, at a money-making business, for years to come, so that I should not be in need of anything." Phelps, in other words, was offered security and cash down in exchange for high risks and possible high profits. Given his temperament, it seems likely that he was glad enough to conclude the deal at the time, although later he must often have kicked himself for doing so. The new contract was signed on 24 December 1850. Soon after this, Singer appointed Phelps to be a traveling agent for the firm. Phelps, as usual, finally acceded, after a certain amount of hesitancy and discontent, to doing something he did not really want to do: "I at first told him I would rather stay at home; but still, as I had bound myself, I should do as they wanted me to."

Zieber remarked, apparently with surprise, that they were able to pay the $1,000 in cash to Phelps "without causing much inconvenience to the firm." But this is really not very surprising, since it was replaced almost at once by a new partner buying in. This was Barzillan Ransom, a manufacturer of cloth bags for packaging salt. He was one of the very early buyers of the machine—in fact he bought the third one to be sold in New York. The unfortunate Gus Singer remembered the sale clearly because, while his father was demonstrating the machine to Mr. Ransom, Gus caught his finger under the plate, with the result that the end of it was taken off by the shuttle driver. This misfortune, however, did not put Mr. Ransom off. He was, on the contrary, so impressed with the working of his machine that he offered to buy Phelps's one-third interest. He was to pay a total of $10,000: three notes, one of $250 and one of $750 at three months, one of $1,000 at four months; and $8,000 to be used as capital in the business.

Zieber managed to cash the three notes, and Singer and Zieber divided the money equally between them. On the strength of this, they opened a New York sales office in a room at the back of Smith and Conant's at 256 Broadway. "The manufacture was still carried on at Boston under the superintendence of a foreman," recalled Singer.

Zieber and Ransom endeavored to sell machines in New York, and I went off traveling for the same purpose, and visited Baltimore, where I sold a few machines, forwarding the proceeds to Boston as fast as the money was collected. Subsequently, in the spring of 1851, I went to Philadelphia with three machines, and had fifteen dollars when I arrived there. I hired a room for an office, paying twelve dollars for the first month's rent in advance, leaving me with but three dollars for expenses. I succeeded in getting money for some machines, and with the money received there succeeded in paying the workmen at Boston and continuing the manufacture.

All this time, said Singer, "I continued the improvement of my machine, forwarding the sketches of the improvements as fast as made, to the foreman of the workshops in Boston."

Ransom, meanwhile, was not proving a satisfactory partner. A short time after joining the firm he became ill, and so could not attend the office regularly. More important, it turned out that the promised capital did not exist. Once more, Singer decided that the unsatisfactory partner would have to go. He seems to have applied very much the same tactics with Ransom as he had used to dislodge Phelps, which was virtually to bully him out of the firm. In March Ransom complained to Phelps, who, one imagines, was only too ready to lend a commiserating ear: "Mr. Singer is rather singular in his views but the writer does not wish to cross him." "Our I. M. Singer assumed so much authority and plays the dictator in such magnificent style that he is perfectly insufferable," he wrote Phelps in April, "and unless he alters his hand promptly we must separate." Such an ultimatum would no doubt have rejoiced Singer, who had just this in mind. Three days later, Ransom wrote Singer, who had gone to Philadelphia without informing the office: "Since receipt of your scolding letter I am in the dark as to your movements . . . I was pained exceedingly by the insulting letters you allowed yourself to send me especially as it was entirely uncalled for." Once more Zieber supplied sympathy from the side.

Singer's extremely irritable disposition, and his abusive and overbearing conduct towards the old gentleman, who had been a down-town merchant and was a man of fine feeling and respectability, soon brought a termination to Mr. Ransom's interest in the business, as well as to his life, probably—for I believe the violence and brutality exhibited towards him very much aggravated his illness [he wrote]. He often used to say "That Singer will be the death of me," and I would see him sob like a child.

Ransom quit the firm in May 1851, Singer and Zieber paying him off with forty sewing machines for his interest. He did indeed die a few months later.

Singer and Zieber were now the sole partners and co-owners of the business, and Zieber would very much have liked to have kept things that way. This, however, was not Singer's idea. He felt that "it was absolutely necessary for us to have in the firm some person of recognized legal ability, who could attend to financial matters and the suits with which we were then threatened." Accordingly, in 1851 he proposed to take into the firm, "without payment of any money upon equal terms with myself and Zieber," a Mr. Edward Clark, a New York lawyer and junior partner in the firm of Jordan, Clark and Company, lawyers whom Singer had previously consulted about his carving machine. The firm was most reputable, Ambrose Jordan, its senior partner and Clark's father-in-law, having recently been made attorney general for the State of New York. It would have been impossible for the penurious firm of I. M. Singer & Co. to afford their fees. The idea was, therefore, that Clark should accept an equal share in the business in return for his legal services. It was not the first time he had entered into such an arrangement with Singer. In October 1850, Singer had transferred to Clark three-eighths of the patent rights on the carving machine, presumably also in payment for services rendered. It was perhaps remarkable that Clark, after such an unprofitable experience, was prepared to try the same thing again. He was, however, destined to do considerably better out of the sewing machine.

Zieber found out about this proposed arrangement in a somewhat unorthodox way. Zieber, it should be recalled, had gone into business with Singer in the first place in the hope of making enough money to clear off some old debts he had acquired in previous business transactions. These debts had not yet been cleared and, now that they were sole owners of I. M. Singer & Co., he felt that he would be in an altogether better position *vis-à-vis* his creditors if he could show them a definite written agreement with Singer as to the terms on which the business was to be operated. "Nothing had yet occurred to interrupt our friendship," he noted, evidently not having let his distress over the cavalier treatment of Phelps and Ransom interfere with the more important aspects of life. "From the creation of two or three sewing machines, which it was at first so difficult to accomplish, our stock in trade had now become worth thousands of dollars, and our prospects in the continuance of the business were unlimited."

Zieber therefore went round to Singer's house—he was now living on East Fifth Street, near Fourth Avenue—on the morning of 10 May to talk about setting their affairs formally in order.

He had not yet risen from his bed and I was requested by Mrs. Singer to go up to his room. After the usual friendly salutations of the morning had passed between us, I spoke to him about the subject in question, upon which he partially raised himself up in his bed, apparently in a great rage, and said in a rough way— usual with him—"What do you mean? By God, you've got enough! You shan't have any more!" He meant that I should have no more of the Phelps interest, about which there had been only a verbal agreement between us . . . I was utterly dumbfounded. If

I had been suddenly condemned to be shot I could not have been more stunned.

It was the revelation of the opportunist where he had expected to find a grateful friend which seems most to have shocked Zieber—and not without reason, although the poor man's faith in human nature seems to have suffered continuous setbacks and by this stage can only be described as more touching than realistic. "I was so overcome," he reported,

> that I sank down upon a chair at the foot of his bed without the power of utterance, not more affected at what I saw I was about to lose than at the unutterable baseness and ingratitude of the man whom I had assisted step by step from the greatest poverty and comparative ignorance in business to the good fortune and prospects he then enjoyed. As soon as I had sufficiently recovered myself I left, without saying anything more.

Zieber now decided he needed help in dealing with his intractable partner.

> I left Singer's house on that May morning, so disagreeable to me, not knowing what I should do and scarcely conscious of which way I was going. As I wandered down the Fourth Avenue, I thought I would go and see Edward Clark . . . He knew Singer, and had been acquainted with our business from the commencement. He had drawn up the agreement between Singer and our deceased partner Ransom, and knew that the money received from the latter had been equally divided between Singer and myself. I supposed him to be an honorable man, and thought that perhaps through his influence with Singer I might obtain justice. I saw Clark and told him what an extremely hard case mine was to be treated in so villainous a manner by Singer, after I had been such a friend to him . . . Clark said that if I had no written agreement with Singer in regard to the third of the business purchased from Phelps, I could not help myself, and reminded me of the "old saw" about "possession being nine points of the law"—thus admitting the advantage Singer was taking of me. Also, in talking about the hold I had on the business according to the original agreement of the three partners in Boston, he said that "although

I was really entitled by that to be a third partner in all the benefits to be derived, Singer might transfer the patent to any one else after it was obtained—that an agreement made about a patent before it was issued was not binding in law." He further remarked that "Singer was a very stubborn and difficult person to get along with," in which opinion I now had good reason to coincide with him entirely. During this interview with Clark, I told him that I was quite disgusted and disheartened with the rascality of Singer, and the advantage which he was then taking of me. That after what he had already done, I believed him capable upon any slight pretext of taking the balance of my interest, and appropriating it to his own use. That I could have no more confidence in him any longer, and that I would sell out, altogether, for a few thousand dollars. To this he coolly replied "perhaps that is the best thing you can do." After further conversation with Clark, he finally made the observation, "Singer will not break the agreement I shall make with him." Hissing the words through his closed teeth in a very spiteful manner, as if he had some tight hold upon his associate in a mean and dishonorable act—for I then immediately discovered that I was, as it were, suspended between the frying-pan and fire—fairly caught between the hawk and buzzard—and understood that Singer, to promote his individual advantage in some way, was about to dispose of a portion of my valuable interest in the business to Clark, who, with full knowledge of all the previous circumstances contained in this statement, was not above receiving it. They were "birds of a feather." From various vague rumors which I then remembered to have heard about some of Singer's previous transactions, there was reason for believing that Clark, who had been his lawyer, had in reality, some ugly hold upon him. And I afterwards noticed that Singer, who quarreled with and bullied almost everyone with whom he had any thing to do, was always very submissive to and apparently afraid of Clark.

Zieber, as usual, was obliged to disguise his finer feelings for the sake of expedience. "June, 1851, Clark was now finally installed a partner in the business, and I was obliged from policy, because I could not do otherwise, to treat him with a certain degree of courtesy and politeness though I could not help despising him in my heart."

The patent at the heart of all this trouble was not finally granted until 12 August 1851. It made three claims, which related to the motion of the shuttle and needle, the control of the thread, and the engagement of the bobbin. The machine was noted by the *Scientific American*, which had taken an interest in sewing machines from the beginning, in their November issue, which gave two detailed drawings and specifications. "This machine does good work," it reported. The rights in the patent, now legally assignable, were divided equally between Singer and Clark. They then presented the enraged Zieber with a paper drawn up by Clark, in which Singer agreed to give Zieber a third of the profits, and Zieber was obliged to give all his personal attention to the business. "This agreement was good enough as far as it went—but it was not what I had a sight to," commented Zieber. He now lived in constant apprehension of his partners' next move, being quite convinced that they were out to deprive him altogether of his share in the business. He received varying advice from the friends to whom he confided his fears, ranging from "I should have killed the damn rascal!" to "Well, Zieber, if I was in your place, I would hold on, just to see what they will do."

Zieber's fears turned out to have been well founded.

On the 15 December 1851, I was suddenly taken ill, and confined to my bed. Our profits at that time, clear of all expenses, amounted to about $25,000. In two or three days Clark came to inquire after my health. He remained five or ten minutes. The next morning Singer came. After entering my room, and asking me how I did, he said to me—"The doctor thinks you won't get over this. Don't you want to give up your interest in the business altogether?" As may readily be imagined, I was very much startled at what he told me and not in a fit state of mind to think much about business. I believed he was telling me the truth, and did not then suspect that it was a trick probably concocted between him and his other partner, to dispose of me, finally, in this fashion as the opportunity offered. As I discovered afterwards, Singer did not know Dr. Anderson, who was attending me, and no such remarks had ever been made by the latter . . . After reflecting for a few minutes I came to the conclusion from what I had previously experienced from my partners, that if anything serious did happen to me, the money I had borrowed from my friends—the very money which had been spent, directly and indirectly, in raising Singer up to the position he now held, would never be paid unless he became legally bound to pay it. So I talked with him about the price. They wanted me to take a very small amount at first. The price was finally fixed at $6,000, and a paper was accordingly written by Clark to that effect, which Singer brought to me the next morning, and I sat up in bed and signed it.

A short time after this Zieber, who was by no means dying, recovered his normal health and realized what he had done. Although he continually vaunted his great previous experience in business, and asserted that he taught Singer all he ever knew about the subject, it may be questioned whether someone so credulous was cut out by nature to be a businessman. At any rate, he was speedily outstripped by his pupil, who seems to have been a shrewd judge of character; Singer seems at any rate very quickly to have assessed who could be browbeaten and who could not.

Edward Clark most certainly could not. It is questionable whether, as Zieber asserts, he had some secret hold over Singer; but there is no disputing that Singer could not do without him and that the business badly needed someone with Clark's particular abilities. Zieber, who hated Clark, thought that "had it not been for what I did and the perseverance and industry of Singer himself, much less would have been accomplished. Great as the business was becoming, much more might have been done, but an illiberal, stringent and cautious policy was now commenced which must have impeded the aggregate gains." In this opinion Zieber was quite alone. Wild optimism, energy and overconfidence were in abundant supply at I. M. Singer & Co.; what was needed was someone with a cool business brain, not over-credulous, and with foresight and planning ability. All these qualities Clark had. There can be no doubt that he drove a good bargain when he took half the patent rights; there can equally be no doubt that Singer got a good deal in his partner. It was largely owing to Clark's astuteness that the firm was able to exploit Singer's patent as effectively as it did.

It was Singer who had brought Clark into the firm; but although it was a commercially astute move, he must have regretted it many times. There was never any disguising the fact that Singer and Clark did not get on. As Zieber put it, "the one as heartily hated his partner as the other, in his turn, despised his fellow . . . There was no personal friendship between them." Indeed, it would have been very surprising if the two men had got on, for two more different characters it would be hard to imagine. Singer was an upstart, brash and crude and frequently blasphemous. Clark, on the other hand, was the scion of an old and respectable family, like Singer's from upstate New York, where for generations they had been settled in Cooperstown. He had been a Sunday School teacher and was correct, lawyerly and shockable. As a business combination they were, however, formidable: Singer's inventiveness, mechanical know-how and boundless drive and energy were ideally complemented by Clark's cool acumen. Neither could do without the other, and so for years they were irretrievably and unwillingly bound together. Singer's transparent bullying had no place in the world of respectable business; it was not a method which could be applied in a large firm, if I. M. Singer & Co. was ever to be such. But with a partner like Clark, there would be no need for it. However, it is easy to see that in 1851, at the beginning of their association, each may have asked himself several times whether he had really got such a good deal as all that.

From Clark's point of view, and despite all Zieber's grand assertions to the contrary when he realized sometime later just what it was he had signed away for $6,000, the business was by no means thriving when he entered it. As recently as April 1851 (Clark finally became a partner in June), all production of machines at Boston was stopped, and all the men working on them discharged, because there was no money with which to keep going. After some weeks, production started again; nevertheless, the firm was hardly as yet solidly established. What Zieber referred to grandly as "our principal offices" consisted of one crowded room at the back of Smith and Conant's. Thomas Jones, the model maker employed by Singer in New York, described it: "There were a couple of chairs and a table, and I nailed up a lid of a box to make a bench in a corner, and got a few tools. That was about all the furniture there was there. There were Mr.

Ransom, Mr. Ransom's son, and Mr. Singer's son, Mr. Singer, Mr. Zieber and myself was there. Mr. Singer's son ran messages etc. I don't believe Mr. Ransom's son was employed there." Business was moving so slowly that, after May 1851, Jones took a machine and went out on the road, through Connecticut and out west. Asked whether his agency was successful, he replied: "Yes, successful at that time. It was a godsend then to sell a machine. It was pretty hard work. Folks couldn't believe that the machines would do, anyhow; and if I sold one machine in two or three weeks, I thought I was doing well."

Indeed, almost everyone seems to have been out on the road at that time. Gus, although he was no more than fourteen, traveled around Baltimore and Washington with his father, exhibiting the machine in stores and hotels; after some weeks of this, he followed his father home to New York, having traveled in charge of a machine by himself for some of the time. (Gus stayed in the office until January 1852, then went on two more selling tours and was finally allowed back to school. By then, the firm was able to afford paid agents.) Yet another Singer son, William, his oldest child by Catharine Haley Singer, was selling in New Jersey and later in the Albany and Troy area. Isaac must have been keeping in touch with his first family; Catharine was now running a boardinghouse on Third Avenue. William's tastes, it seems, were too like his father's to be altogether good for business. Ransom complained to Singer in one letter that he didn't know where William was to be found, and that "the last two days he spent in the office he was engaged in writing a play for one of the theaters."

This was the organization—if such it could be called—to which Clark was now committed. Singer, for his part, soon began to grumble about his new partner. He complained to Zieber that Clark "had brought little or no money into the business—he was not a good merchant—he had no energy for such a business—etc. A moderate salary would have procured as good a man for our purpose." Zieber, no doubt, was all too ready to agree.

However, Clark was just about to come into his own, for it was at this point that the figure of Elias Howe once again loomed upon the horizon. For a second time he presented himself at Singer's with a demand that some settlement of his patent claims be arrived at: otherwise he would bring suit against them and effectively prevent the manufacture or sale of any more Singer machines. The price Howe was now demanding for rights to manufacture under his patent was $25,000. Once again he was shown the door.

This turned out to be a very grave mistake on the part of Singer and Clark, and one which would eventually cost them a great deal more than $25,000. In fact at this stage both Singer and Clark seriously underestimated Howe. In a letter he wrote in 1852 to Singer's Boston agent, Clark said: "Howe is a perfect humbug. He knows quite well he never invented anything of value. We have sued him for saying that he is entitled exclusively to use of the combination of needle and shuttle . . . He has never dared to make any attack on us and never will."

In this assessment Clark was quite wrong. Howe was suing every manufacturer of sewing machines he could find, and I. M. Singer & Co. were to be no exception. In 1851 there were six manufacturers in the sewing machine business and by 1853 the four biggest, apart from Singer, had all accepted Howe's terms and were manufacturing under his license. This put Singer and Clark in a very difficult situation. Their own business was dogged by threats of litigation, with all the risk and expense entailed. Those of their rivals were not. But they had taken a stance which made it unthinkable to settle with Howe, even if Howe, who was now negotiating from a position of strength, had still been willing to do so. For the next three years, therefore, all the firm's profits, and most of Edward Clark's energies, were devoted to the legal battle with Howe and the other manufacturers—a battle which was soon known by all the newspapers as the "Sewing Machine War."

★ ★ ★

The "Sewing Machine War" was conducted on two fronts: in the newspapers and in the courts. The weapons for the newspaper battle were advertisements. The classified columns which formed so important a part of nineteenth-century newspapers were filled with verbal slingshot aimed at boosting the advertiser while sullying the name of his rivals. Thus Howe advertised on 29 July 1853:

The Sewing Machine—It has been recently decided by the United States Court that Elias Howe, Jr., of No. 305 Broadway, was the originator of the Sewing Machines now extensively used. Call at his office and see forty of them in constant use upon cloth, leather, etc., and judge for yourselves as to their practicability. Also see a certified copy, from the records of the United States Court, of the injunction against Singer's machine (so called) which is conclusive. He has a suit now pending against the two-needle machines, so called, and is about commencing suit against all others offered to the public except those licensed under his patent. You that want sewing machines, be cautious how you purchase them of others than him or those licensed under him, else the law will compel you to pay twice over.

The very next column carried an advertisement from I. M. Singer & Co.:

Sewing Machines.—For the last two years Elias Howe, Jr., of Massachusetts, has been threatening suits and injunctions against all the world who make, use or sell Sewing Machines—claiming himself to be the original inventor thereof. We have sold many machines—are selling them rapidly, and have good right to sell them. The public do not acknowledge Mr. Howe's pretensions, and for the best of reasons. 1. Machines made according to Howe's patent are of no practical use. He tried several years without being able to introduce one. 2. It is notorious, especially in New-York, that Howe was not the original inventor of the machine combining the needle and shuttle, and that his claim to that is not valid . . . Finally—We make and sell the best SEWING MACHINES—the only good ones in use; and so far from being under injunction, we are ready to sell and have perfect right to sell our Straight Needle, Perpendicular Action Sewing Machines —secured by two patents in the United States—at the very low price of $100, at our offices in New-York, Boston, Philadelphia, Baltimore and Cincinnati.

"CAUTION," thundered Howe.

ALL PERSONS ARE CAUTIONED against publishing the libelous advertisements of I. M. Singer & Co. against me as they will be prosecuted to the fullest extent of the law for such publications. I have this day commenced action for libel against the publishers of the said Singer & Co.'s infamous libel upon me in this morning's *Tribune* . . .

People's opinions regarding the issues in the Sewing Machine War varied enormously, depending, of course, on which side they were on. The affiliations of the author of the following effusion, for instance, are apparent:

From the outset, Singer & Co. resisted, at great expense, the demands and pretensions of Howe, fighting single-handed the battle of the inventors and the great world which was waiting for cheap machines. Howe was endeavoring to establish a monopoly, strong and compact, which meant dear machines to the weary-fingered women who were still singing the dreary "Song of the Shirt": Singer & Co. were struggling to throw the business open to fair and honest competition at moderate prices. All the other manufacturers had yielded to Howe at the first, and were conducting their business without interruption under his licenses. They viewed the contest between Howe and Singer & Co. much as the traditional frontiersman's wife regarded a terrible struggle between her husband and a grizzly, merely remarking that "it didn't make much odds to her which won, but she allus loved to see a right lively fight."

The combatants raged against each other not only in public but in private. Singer, according to Zieber,

raved to put his foot upon the neck of Howe . . . After the deaths of Arrowsmith and Bliss, Sr. [Howe's financial backer] who were interested as witnesses, I believe, in the law-suits which had taken place, Singer triumphantly exclaimed that "he firmly believed they were burning in hell for the perjuries they had committed to deprive him of his rights." Wheeler and Wilson, he said, "were trying to rob him," and Potter of the Grover and Baker Co., "was a damn scoundrel." He cursed the *Scientific American*, because "it did not do him justice"—and Willcox and Gibbs caused him to pass many unhappy hours. When young Bliss [who had inherited some interest in the Howe patent] died from the effects of a fall from his horse, Singer said "it was the providence of God avenging his Senior's wrongs."

Singer and Clark might, according to viewpoint, be the very St. Georges of the sewing machine world, or they might be the dragon itself; but mere opinion was of no account unless and until supported by a suitable judgment in law. Clark might asseverate that Howe was a humbug, but could the same thing be said about his patent rights? It was impossible merely to ignore him, because no sewing machine could work without incorporating at least three of Howe's patented innovations. Accordingly, Singer and Clark's only legal hope lay in proving that Howe had not in fact been the first person to invent a machine incorporating these features. If they could prove this, Howe's patent would be invalid and he, too, could be said to be in infringement. Their first effort in this direction was to try to prove that the Chinese had invented a workable sewing machine centuries earlier. It was a game attempt—the Chinese had indeed invented most things at one time or another—but it failed. However, it did not take long before their assiduous inquiries turned up the story of old Walter Hunt and his 1834 machine.

Walter Hunt was still in business, working now for a small machine shop in an alley off Abingdon Square. How Singer and Clark must have gloated when they heard his story! It was a godsend, the stuff of their wildest hopes. Walter Hunt had made, or tried to make, his sewing machine as early as 1834. It had worked, if imperfectly, as we have seen; but perfection was not what they were interested in, only the details of the mechanism. Howe's machine had not worked perfectly, come to that, but he was still a very effective thorn in their side. The really important feature of Hunt's machine, from the infringers' point of view, was that it had used a lock stitch which was formed with a shuttle. There it was: Howe had done no more than re-create something Walter Hunt had invented twelve years earlier.

Nobody, however, was going to take Walter Hunt's word for this on trust. Too much was at stake. It was essential that at least one of Hunt's old machines be found and proved workable; the courts could take it from there. But all this had taken place nearly twenty years earlier and Hunt had never been the kind of man who keeps everything neatly filed away in boxes against the day when it might come in useful again. Indeed, his whole character worked in quite the opposite way. Money-making had never been his prime objective. He was only interested in whatever idea happened to occupy the forefront of his mind at any given moment—and such was the fertility of his invention that there almost always *was* something new. Old ideas were finished, explored, and retained no further interest for him. He couldn't imagine what might have happened to his old machine.

Singer was not the kind of man to let a subject drop so easily. He was interested in money-making—"the dimes, not the invention"— and it seemed as though the finding of Hunt's machine was the first prerequisite to the dimes rolling in. Day after day he returned to Hunt to try to jog his memory. Where had he been working when he made his machine? What had he done with his model? Well, thought Hunt, it was just possible that George Arrowsmith might have kept a model in his shop in Gold Street . . . In the attic of the Gold Street house, they eventually found some rusty and broken pieces of metal which Hunt declared were the remains of his original machine. Could he take these mangled pieces and put them together to complete the machine as it was originally made and get it to sew? Hunt thought he could. Then he must try! Singer supplied him with money and encouragement and urged on him the possibility of a fortune to be made, *if only* he could get his machine to work. But try as he might, and despite the constant efforts of would-be helpers, Hunt could not transform his rusty bits of old iron into a working sewing machine. What was to be done? A supercilious lawyer in a patent suit some years later evoked the scene:

> William Whiting, of Boston, was among the advisers, and his penetrating ingenuity and anatomical experience was brought to bear upon the parts of the old carcass, which had been found in the attic in Gold Street, and to him remained the honor of doing in retirement what could not be done in assembled council. He, after the lapse of many days, informed Mr. Hunt what he might have done, and Mr. Hunt . . . agreed, and subsequently insisted, that *that was just what he did do.*

Walter Hunt now belatedly applied for a patent on his machine, and on 19 September 1853 he put an announcement in the *New York Tribune*:

TO THE PUBLIC—I perceive that Elias Howe, Jr., is advertising himself as patentee of the Original Sewing Machine, and claiming that all who use machines having a needle or needles with an eye near the point, are responsible to him. These statements I contradict . . . Howe was not the original and first inventor of the machine on which he obtained his patent. He did not invent the needle with the eye near the point. He was not the original inventor of the combination of the eye-pointed needle and the shuttle, making the interlocked stitch with two threads, now in common use. These things, which form the essential basis of all Sewing Machines, were first invented by me, and were combined in good operative Sewing Machines which were used and extensively exhibited, both in New York and Baltimore, more than ten years before Howe's patent was granted . . . I have taken measures as soon as adverse circumstances could permit, to enforce my rights by applying for a patent for my original invention. I am by law entitled to it, and in due course no doubt will get it.

The *Scientific American* was very disapproving of this step. "We take a positive position in opposition to the claims and assumptions set up in this card," it declared, and gave its reasons:

Mr. Hunt *may* have invented what he claims, but at this date, when the value of such machines have been brought into public notice by others, and seven years after Howe obtained his patent, it has rather an ugly appearance to set up ten years' prior claims to the lock stitch and eye-pointed needle. Since the time when it is asserted he invented his machine, he found means to obtain patents, and to induce others to purchase inventions of far less importance and value; how came this one to be neglected? We are opposed to such rusty claims, especially by one so well versed in patents and inventions . . . We want to see the means provided by law to settle such controversies with dispatch, in order that the ear of the public may not be used as a kettle drum on which to beat the loudest tones for personal purposes.

The case of Hunt *vs.* Howe was finally brought to trial on 24 May 1854. Charles Mason, Commissioner of the Patent Office, in his testimony took very much the same view as had the *Scientific American*:

When the first inventor allows his discovery to slumber for eighteen years, with no probability of its ever being brought into useful activity, and when it is only resurrected to supplant and strangle an invention which has been given to the public, and which has been made practically useful, all reasonable presumption should be in favor of the inventor who has been the means of conferring the real benefit upon the world.

Encouraged, Howe now brought a suit in Boston to restrain two firms from selling Singer machines. On 1 July Judge Sprague gave his opinion in this case: "The plaintiff's patent is valid. Other machines are infringements . . . There is no evidence in this case that leaves a shadow of doubt that, for all the benefit conferred upon the public by the introduction of a sewing machine, the public is indebted to Mr. Howe." Singer was ordered to pay Howe $15,000. Howe now threatened to sue Singer in New York and New Jersey as well, and it was obvious that he would succeed. As one pro-Singer article succinctly put it, Singer "had a dislike to bringing it before a jury in New York"—for the very good reason that he was sure to lose. A settlement was therefore negotiated. Singer, no doubt gnashing his teeth, finally capitulated and agreed to manufacture under license from Howe, who took a twenty-five-dollar royalty on every machine sold.

At last, nine years after he had first patented his machine with such high hopes, Elias Howe's dreams of a fortune looked like being realized. He bought out his partner, George Bliss, and became for the first time sole proprietor of his patent, "just when it was about to yield a princely revenue," as his biographer observed. The most he had ever earned from his machine until now was a few hundred dollars a year. Now his income increased until eventually he was earning well over $200,000 a year. In 1860, he applied for an extension to his patent and by 1867, when the patent finally expired, it was calculated that his earnings were not far short of $2 million. The capitalist was proved right in the end.

As soon as this point was finally settled and out of the way, all the sewing machine manufacturers got busily down to the job of suing each other out of existence. The great advantage of the sewing machine, from the lawyers' point of view, was that, as we have seen, no one complete and entire working sewing machine was ever invented by one person unaided. Conversely, every sewing machine manufactured had, perforce, to contain parts which had been invented and patented by several different people. By 1856, I. M. Singer & Co. alone controlled twenty-five different patents, the original inventions of nine different people apart from Singer. By 1867, nearly nine hundred patents had been issued on improvements to sewing machines, although, as a contemporary commented, "Perhaps thirty of these patents are valuable; but the great improvements are not more than ten in number, and most of these were made in the infancy of the machine." Judge Sprague's decision had effectively given Elias Howe control of the entire market—all the other manufacturers who had considered standing out against him were more or less forced to take out one of his licenses. But they in turn lost no opportunity to sue anyone they could, including Howe himself when he set up a factory of his own, for infringement of any patent they happened to own. The resulting legal free-for-all may not have produced many sewing machines, but it gave rise to immense amounts of paper. "We the other day looked over the testimony taken in one of the suits which Messrs. Grover and Baker have had to sustain in defense of their well-known 'stitch,'" wrote James Parton in 1867.

> The testimony in that single case fills two immense volumes, containing three thousand five hundred and seventy-five pages. At the Wheeler and Wilson establishment in Broadway, there is a library of similar volumes, resembling in appearance a quantity of London and Paris Directories.—The Singer Company are equally blessed with sewing machine literature, and Mr. Howe has chests full of it. We learn from these volumes that there is no useful device connected with the apparatus, the invention of which is not claimed by more than one person.

Orlando Potter, the president of the Grover and Baker Co., estimated that in the years 1855–56,

> He [Singer] had suits pending in Philadelphia—several of them—some against Wheeler and Wilson, and some against Grover and Baker; he had fifteen or sixteen suits pending in the Northern District of New York, against Wheeler and Wilson and Grover

and Baker—most of them against Grover and Baker, and several in the Southern District of New York, against Wheeler and Wilson and Grover and Baker and against one made by Bartholf...

Such endless litigation consumed not only everyone's profits but also their time. Enormous care and effort had to be expended, first to chase up the relevant witnesses, and then to make sure that their evidence was all that it ought to have been. Such organization fell to the part of Clark. In 1855, for example, we find him writing:

In some suits we have commenced in Philadelphia against the Grover and Baker and the Wheeler and Willcox [sic] machines, a conviction is about to be issued to examine Mr. Charles Morey as a witness. I shall name Mr. Fleischsmann [a patents agent] as commissioner and hope he will attend to see that Morey testifies correctly. I am much afraid of the sinister influence of Mr. E. Baker with Morey. Fortunately however Morey's interest is to swear to the truth, and just as I would have him ...

The situation was an absurd one and it was obviously in no one's interest that it should continue, except that of Elias Howe who continued to collect twenty-five dollars on every machine sold. It came to a head in October 1856, when Singer, Wheeler and Wilson, and Grover and Baker all met in Albany, when the cases in which each accused the other of infringement were to be tried. A good three months of acrimony in court lay ahead of them. Nobody viewed the prospect with much enthusiasm, but everybody was far too indignant about his own interests to spare time to consider the interests of the whole.

The one exception to this blinkered outlook was Orlando B. Potter, the lawyer who was president of the Grover and Baker Company. Perhaps supreme boredom at the prospect of the next three months fostered a certain detachment in him. However that may be, Orlando Potter now had one of those ideas which appear in retrospect to be so simple, the perfect and obvious solution to the problem, but which are so hard to imagine for the first time. He pointed out that here, gathered together, were most of the people who controlled the manufacture of sewing machines. Why, instead of each pursuing his own factional interests to the ultimate benefit of no one, should they not combine? They owned, between them, almost all the patents worth owning. Why not pool their interests instead of wasting time and money conducting these interminable fights?

What Orlando Potter had devised was the first patent pool, and it was in fact the only way that anyone would ever be able to produce complicated pieces of machinery without tying themselves up in legal knots over patent rights. Later on, patent pools were used in the manufacture of automobiles, radios and other goods. The terms of the agreement were simple. All the interested parties were to pool any patents they held. For a fee of fifteen dollars on every machine sold they agreed to license any member of the Sewing Machine Combination, as they called themselves, to use any device or devices belonging to them. Part of this money was to be reserved to fight infringers, and the rest would be divided between them.

However, the Combination would only work if Elias Howe could be persuaded to join it, which naturally he was very unwilling to do: where he held sole control what had he to gain from ceding to joint control? In the end, though, persuaded by the prospects of increased trade, he agreed to join the others provided he could retain favorable terms. It was accordingly agreed that over and above his share Howe was to receive five dollars on every machine sold in the United States and one dollar on every machine exported. Howe also stipulated that at least twenty-four manufacturers should be licensed members of the Combination to prevent the formation of a monopoly.

Howe had of course put his finger on the great danger inherent in such an organization, for, once one group of traders establishes itself in a position of such strength, it is then able to kill off any independent opposition and may effectively dictate the terms of trade. Thus, while he may have cut down on litigation in the short run among the members of his Combination, Orlando Potter also provided good livings for generations of lawyers to come and gave rise to whole government departments. The Anti-Trust Division of the United States Department of Justice was formed very largely to deal with the fruits of Mr. Potter's brainwave.

Indeed, despite Elias Howe's good intentions, it did not take long before the press realized that the workings of such organizations as the Sewing Machine Combination might not be altogether in the public interest. In 1860 Howe applied to the Commissioner of Patents for an extension of his patent for another seven years over and above the normal fourteen years which had then elapsed. He claimed that his profits from the machine—he had by then earned $468,000 from it —were not commensurate with the value of his invention. This application was, as we have seen, granted; but public opinion was very much against the decision. "It is not within the history of invention," stormed Horace Greeley's *New York Daily Tribune*,

> that any such sum was ever before received by an inventor. Without a dollar's capital, without risking a cent in the contingencies of business, without the cares and toil of manufacturing, or the anxieties of commercialism, but because he happened to be the convenient tool of a band of monopolists, he has quietly sat still and had poured in upon him this enormous wealth, such as few men, with a lifetime of toil, ever attain to. From what source are all these large sums gathered? . . . The principal sale of sewing machines is to the poor needle-women, widows and orphans, those who by toiling day and night barely gain the bread for starving relatives and themselves. From each of these poor needle-women has this patentee drawn not less than five dollars; but more, and more still, the cost of making these machines may be set down as varying between five and twenty-five dollars. The *cheapest* sold by the parties to this most odious monopoly is *fifty* dollars.

This, however, was still in the future. The immediate result was the agreement which made it possible for the manufacturers to get down to the business of making sewing machines. This they now did with gusto. In 1856 I. M. Singer & Co. manufactured 2,564 machines. In 1860 they manufactured 13,000. The age of mass production had arrived.

6

The Beginning of Mass Production

The creation of the Sewing Machine Combination opened the way to
something as important in its way as the original invention of the
machine itself. The sewing machine was the first of a whole new class
of domestic appliances. It was something which every household
might potentially want, but which, unlike other items in such
potentially great demand at the time, could not be made anywhere
other than in the few sewing machine factories. But this in itself was a
significant curb on demand: if sewing machines had been as popular in
1853 as they were in 1863, the factories could not possibly have coped.
However, demand was not high in the early years and one great
deterrent was the exceedingly high price of the item.

The Singer Company's historian, John Scott, declared that when
they contested Howe's patent, "Singer & Co. were struggling to
throw the business open to fair competition at moderate prices." But
there are several reasons why it seems reasonable to doubt this
generous interpretation of their motives. Sheer obstinacy and
defiance were far more characteristic of Singer than any concern for
public well-being—and it was not yet in the power of Singer or anyone else to reduce prices if he also wanted to remain in business. Even
if Howe had never pressed his claims, it is most unlikely that the price
of sewing machines could have been brought down to a more reasonable level much before this did in fact happen at the end of the decade,
because the techniques which made this finally possible were not yet
fully developed.

The great transition was made between the years 1858 and 1859. In 1858, the firm of Wheeler and Wilson made 7,978 machines; Grover and Baker, 5,070; I. M. Singer & Co., 3,591. In 1859, Wheeler and Wilson made 21,306; Grover and Baker, 10,280; and I. M. Singer & Co., 10,038. By 1870, the Singer Manufacturing Company (as it by then was) reached six figures in annual production for the first time, manufacturing 127,833 machines (in 1867 it had taken over from Wheeler and Wilson as the biggest manufacturer). Mass production had well and truly arrived, and its effect on the price of individual machines was, as might be expected, dramatic. Singer machines in 1850 cost $125 each. By 1870, the average price of a sewing machine was $64—a substantial profit on a production cost of $12 per machine. This had been achieved by the introduction of a production-line system using interchangeable parts, the first time such a system was ever applied in general manufacture.

A description of the first Singer factory in New York gives a clear idea of why early sewing machines had to cost so much. This "factory" was a room 25 feet by 50 on Center Street, over what was then the New Haven Railroad Depot. Almost all the parts of each machine were produced by handwork at the bench. This meant that only skilled craftsmen could be employed, necessitating a high wage bill. It also meant that no two machines, or any of their component parts, were exactly similar either in shape or in the way they fitted together. The consequences of this system of production were several, all highly uncommercial. The first was that machines were very slow to produce: it would not have taken a particularly large order to swamp production facilities entirely. Second, because of the time taken in production and the quality of labor needed, machines had to be expensive. Third, no sewing machine could be a very reliable tool if removed very far from a workshop capable of replacing broken parts —which might well mean making them from scratch: there was no concept of "spares," since there was no way of guaranteeing that they would fit. With the introduction of interchangeable parts, all of these difficulties were at once overcome.

The key to a successful system for manufacturing complex machines with interchangeable parts is accurate measurement, not just to the sixteenth or twentieth of an inch achievable by a meticulous craftsman, but to the thousandth or ten-thousandth of an inch achievable only by a machine. The system depends on access to accurate machine tools and gauges. Until such tools were available, the sewing machine, along with virtually all machinery other than firearms, remained a crafted rather than a manufactured object.

The armaments industry has always been a pioneer of new engineering and industrial techniques. In our own day its by-products vary from fireproof ceramics to photovoltaic solar cells. In the nineteenth century as today, governments were more ready to underwrite research and development in this area than in any other. New techniques take a lot of time and money to perfect, and they need the certainty of large markets and assured orders if they are to be economically worthwhile. Governments are usually the surest source of such certainties; and what governments always need, or at any rate want, are more advanced weapons. In the eighteenth century, the accurate boring machines which eventually made possible the production of cylinders for steam engines were first developed in order to bore cannon. In the nineteenth century, the system of interchangeable parts which was eventually to result in the production-line manufacture of such machines as the typewriter, the sewing machine and the automobile was developed from techniques introduced early in the century to ensure an adequate supply of muskets to the American soldiery in the War of 1812. The fact that mass-production techniques were first developed in the United States rather than in Europe is a direct consequence of the fact that the American army relied on muskets rather than cannon.

The main reason for this was that whereas in Europe the chief use of guns was in warfare, in pioneer America the gun was an essential tool of the frontiersman. The settler's gun was his source both of food and self-defense as he labored to clear his ground and establish his crops in the face of marauding bears and attacking Indians. Americans, therefore, needed more light guns than had ever been made before. Moreover, the sort of gun which had to be repaired by a skilled gunsmith every time it went wrong was no use to the pioneer. He might be a hundred miles or more from the nearest settlement, and there was no guarantee that there would be a gunsmith in residence when he got there. The need was then not just for more reliable and accurate light weapons, but also for guns which were easily repaired. The use of interchangeable parts would make it possible to repair guns from stocks of spares which could be carried by any store, and to reuse parts of an old gun which was in some way irretrievably damaged.

Thomas Jefferson was the first American to report on the possibilities of interchangeable parts in muskets. He encountered them in 1785 in France, where the engineer Le Blanc pioneered the system. When Jefferson visited Le Blanc's workshop, the engineer handed him a box containing parts of fifty musket locks arranged in compartments. "I put several of them together myself taking pieces at hazard as they came to hand," reported Jefferson, "and they fitted in the most perfect manner. The advantages of this, when arms need repair, are evident. He effects it by tools of his own contrivance, which at the same time abridge the work, so that he thinks he shall be able to furnish the muskets two livres cheaper than the common price..."

The key phrase is "tools of his own contrivance." We have seen that the accuracy needed for true interchangeability can be guaranteed only by a machine. What is more, the machine has to be designed specifically for the job which it is to perform. There is no need for it to do any other job, but it must do this one perfectly. And—just as important—all the machines making different parts for a particular piece of equipment must be designed to one standard gauge. A screw has not just to fit its particular socket perfectly: it must fit any similar socket perfectly. Therefore it is not just the design of the piece of equipment—in this case, the musket—which is important. The design of the machines which are to manufacture it is perhaps even more important. By 1811, the United States was reaping the benefits of Jefferson's observations. In that year, Captain John H. Hall, engineer to the government armory at Springfield, Massachusetts, patented his rifle and the machinery to manufacture it on the interchangeable system. In the same year, the British government had on hand no less than 200,000 musket barrels which were useless because of the lack of skilled men to repair the locks.

The report of a visitor to Eli Whitney's armory in the early 1800s gives a picture of the kind of production line which was already established at that early date:

> The several parts of the muskets were, under this system, carried along through the various processes of manufacture, in lots of some hundreds or thousands of each. In their various stages of progress, they were made to undergo successive operations by machinery, which not only vastly abridged the labor, but at the same time so fixed and determined their form and dimensions, as to make comparatively little skill necessary in the manual operations. Such were the construction and arrangement of this machinery that it could be worked by persons of little or no experience, and yet it performed the work with so much precision, that when, in the latter stages of the process, the several parts of the musket came to be put together, they were as readily adapted to each other as if each had been made for its respective fellow.

Such a process was obviously ideal for the American labor market. Many of the conditions already discussed in connection with the invention of the sewing machine are relevant here. American workers and manufacturers welcomed new manufacturing processes in the same way and for the same reasons as they welcomed new labor-saving machines. The shortage of skilled labor in the United States meant that craftsmen did not feel their jobs or prestige threatened. The new processes meant more job opportunities for unskilled laborers, who were thus able to earn good money, since it was the most progressive and successful manufacturers who introduced the new machines, and they paid high wages. Once again, American inventiveness and independence were demonstrated. The stupid British edict of 1785, which forbade, under threat of heavy penalties, the export to America of any tool, machine or engine, or the emigration of any individual in any way connected with the iron industry or manufacturing trades associated with it, had exactly the opposite effect to that intended. The idea had, of course, been to ensure that the United States remained permanently dependent on British expertise for her manufactured goods, so that Britain could maintain both a tame market and a permanent source of raw materials. The effect was, naturally, to encourage native American inventiveness in this area. The development of the interchangeable-parts system in America marked the great break with British engineering tradition. The differences become apparent if one compares an early Whitney machine from the armory with any contemporary English machine tool. Each was the very latest thing in its time and place. The American machines are relatively small and designed with one particular job in mind. So long as they performed that perfectly, it did not matter that they could do nothing else. The English machines, on the other hand, are much more massive, built to last a lifetime and to serve any number of different purposes: they were built for maximum serviceability and adaptability.

For Singer and his co-manufacturers, all this meant that they were on the right side of the Atlantic to do their job most successfully. The system of interchangeable parts had been far more fully developed by the middle of the nineteenth century in the United States than anywhere in Europe. This in turn meant that the United States was the only place where it was then possible to bring a complex mechanism such as the sewing machine into successful commercial production. Only such a system could ensure the precision (for "fine sewing") and durability (to survive handling by unskilled members of the public) which were essential to its successful manufacture.

This being so, the obvious question is why they waited until the end of the decade before introducing the new methods? Why persevere with inefficient and laborious hand-crafting when machines were available which could do the job so much better?

First, and most simply, all the machinery was *not* available when Howe and Singer first developed their machines. The key tool in the automatic manufacture of sewing machines was the milling machine. Shapers, planers and lathes there were—Orson Phelps probably had a fine selection of them in his machine shop. But work which can be done on a shaper, planer or lathe may often be performed to better effect by milling—provided there is enough repetition of the work to justify making a milling cutter of the correct shape and size. The milling machine has two advantages over the other tools: the large number of cutting edges allows for high-speed machining without overheating, and particular shapes can be obtained in a single operation by the use of formed cutters. Despite this, the use of milling cutters was very restricted in the early days of machine tools, because such cutters were very difficult to make and to keep sharp. Eli Whitney made a small milling machine; but the first milling machine to be manufactured for sale was not designed until 1848. The period 1848–52 also saw the production of the first accurate profile-cutting machine. The first product which can really be said to have been manufactured entirely on specialized machinery was the pistol manufactured by Colonel Samuel Colt, who installed in his armory, during the years 1853–55, 1,400 specially designed machines, including die-stamping machines and drop hammers for the rapid production of very accurate small forgings, special jigs, fixtures and gauges. It so happened that the construction of the sewing machine very much suited the methods which had been evolved by the armories, with the result that the two trades to some extent became associated with each other. The sewing machine manufacturers adapted the armorers' machines, and several New England armories in fact manufactured sewing machines during the 1850s—though they gave this up for obvious reasons during the Civil War. The manufacture of the sewing machine fully extended the very latest developments in machine-tool making of its time.

Very much connected with this was the availability of investment funds. Bearing this in mind, the timing of the great expansion in sewing machine production is absolutely logical. The equipping of a large factory with all the latest purpose-designed machines was and is a very expensive undertaking. Nobody was going to be so foolhardy as to commit himself to such an expense while the terms of trade were as uncertain as they must have seemed to be before the creation of the Combination. Before the 1856 settlement all the companies knew for certain that they would have to devote a considerable proportion of their funds and energies each year to litigation. After 1856, they could feel free to get on with the job in hand—and they immediately set about it in magnificent manner, each company trying to outdo its rivals. Naturally, as soon as one firm adopted the new production methods all were virtually forced to do so, or be out-priced and out-produced.

Thus by December 1857, I. M. Singer & Co. were well ahead with the construction of a new factory of the most modern design on New York's Mott Street. Their house journal, *I. M. Singer & Co.'s Gazette*, described it in rapturous detail:

> On four lots of ground, on the easterly side of Mott Street, between Broome and Spring Streets, in about the most central part of the city of New York we are now erecting a fire-proof building for a machine-shop and iron-foundry, which, when completed, will be the [finest] edifice, without any exception, ever built for manufacturing purposes. It is to be constructed of materials indestructible by fire. This building will be six stories in height, besides basement and cellar, making eight floors in the whole. The cast-iron front on Mott Street, will be of an elegant design and highly ornamented, so that few buildings in New York or elsewhere can lay claim to more architectural beauty. Messrs. D. D. Badger & Co., the eminent builders in iron, have the contract to erect this manufactory, and are to receive one hundred and twenty thousand dollars for the work. It is estimated that when this building shall be completed, and fully stocked with tools and machinery for the manufacture of our Sewing Machines, that the aggregate value of the establishment will be not less than three hundred thousand dollars . . . Our new iron-foundry, which is connected with our manufactory, is already completed and in operation.

If the tone seems to imply that the factory and its accouterments were even more of an innovation than the machine they were being built to manufacture, this is perfectly logical: indeed they were. This was the very first time that such methods had been applied and such a huge investment made to manufacture anything other than guns. The sewing machine manufacturers were breaking new ground and risking huge sums of money to do so; they wanted to make quite sure that the world realized what was going on and was duly impressed.

In this they were not disappointed. The wonders of the new factories provided copy for most of the Eastern press at one time or another during the next few years. A detailed description, floor by floor, of the Singer factory, taken from the Singer *Gazette*, was reprinted verbatim by both the Philadelphia *Daily Press* and *Harper's* "Editor's Table" in 1860. In 1863, the New York *Daily Tribune* devoted an entire page to the wonders of the Wheeler and Wilson factory, in a tone of childlike amazement at the extreme modernity of the new methods:

Let us suppose a load of pig-iron placed upon a car at the Lehigh, Pa., mines [wrote the *Tribune*] and run directly into the works, which are connected by a side track with the New York and New Haven Railroad, and from that into the furnace, thence into the molds which cover the floor of one immense room, and employ a score of men.

From the foundry the castings are taken upon small railways to the first floor of the finishing shop, which is a three-story building, 36 feet wide and 550 feet long, with lines of shafting from end to end of each floor, which drive two or three tiers of machines upon the center of the floor and lines along each well-lighted side . . .

HOW THE CAST IRON IS PLANED AND BORED: This was a tiresome job when all was done by hand; when each part of each machine was fitted to its fellows by cutting and filing, and when the parts of two machines must be kept separate. Now all are so exactly alike that a thousand pieces are finished and thrown into a box together, and each one forming a part of a machine, and never requiring the stroke of a file to adjust it, though the parts might be a thousand miles away from each other.

Here is a man just taking a bed plate from a great pile just unloaded from the foundry . . . To insure perfect accuracy these plates—as well as every other part, in all their changes from one machine to another, for all the cutting and boring—are fixed upon hardened steel guides, so that it is impossible for one part to be made amiss, or a hole to be bored wrong because, before it can begin to bore, the drill most pass through the guide hole in the steel plate.

Each machine is fitted to do one kind of work and no other. Thus, as the bed plate is planed upon the bottom in one, it *is* planed on the top in another, and another cuts away such parts of the metal as may be necessary, and another bores a certain portion of the holes, and so it is passed along . . . Even the small screws are made upon machines that would cost $100 each.

Such factories were to be found nowhere else in the world. The English *Mechanic's Journal*, for instance, thought fit to explain to its readers that "in the United States, the manufacture of sewing-machines is carried on by improved machinery, in large factories specially devoted to the purpose, and consequently the machines themselves are, as pieces of mechanism, of the most perfect construction."

The great sewing machine factories indeed marked an important stage in the relative positions of American and British engineering. English manufacturers had long assumed that their equipment was superior to anything made elsewhere, and for many years this assumption was perfectly justified. The first seepings of doubt began to creep in at the Great Exhibition of 1851. When Cyrus McCormick's reaper was shown there, the London *Times* ridiculed it.* But when it was demonstrated, the reaper won a prize and the *Times* apologized.

* The *Times*'s attitude to American engineering at this time may be seen in a note it reprinted from *Punch* on 22 May 1851: "American Contributions to the Great Exhibition.—All American contributions have not yet arrived. This delay accounts for the empty space, which is so painfully noticed, at the end of the Exhibition. The following rare goods, however, are on their way from the United States, and may partially make up for the lamentable American deficiency:
The Leg of a Multiplication Table.
The Pair of Ship's Stays found drying on the Equinoctial Line.
The Bonnet generally worn with the Veil of Nature.
The Paletot of the Heavy Swell of the Atlantic.
The Key of Locke's Music.
A Tooth-brush for the Mouth of the Thames.
A Hat-stand made out of the Horns of a Dilemma.
A Tumbler for the Jug of a Nightingale.
A Mattress for the Falls of Niagara.
A Dressing-case for the Mirror of Nature.
The Whip with which America flogs all Creation—especially the coloured portion of it. And, lastly—
The Tremendous Wooden Style that separates the American from the English fields of literature."

Indeed, the examples of American engineering so impressed the leaders of British industry that in 1853 they sent the famous engineer and machine-tool maker, Joseph Whitworth, to the United States to see what examples might usefully be followed in Britain—a curious piece of irony when you consider the benighted British policy which had given such a boost to American engineering inventiveness only seventy years before. Whitworth's astonished reactions to the American workmen's receptivity to progress and innovation have already been noted. In the United States, he reported, "the workmen hail with satisfaction all mechanical improvements, the importance and value of which, as releasing them from the drudgery of unskilled labour, they are enabled by education to understand and appreciate."

One may quarrel with the conclusion he drew—the introduction of machines with built-in skills was the very last thing to add to the intrinsic interest of the job, and it introduced, precisely, the "drudgery of unskilled labour" to replace the slower and costlier, but more interesting, skilled labor—but there is no quarreling with the fact that American workmen welcomed machines where the British did not. They were working for the cash which could buy independence, and machines meant more money. Whitworth was speaking as a man with strong vested interests in the wider introduction of machines. Machines were his life; he designed and built them. The more machines employers bought, the happier he was. But in England, although farsighted employers might want to install the latest machine, the work force did not and fought against it. Luckily for Singer, he was based in America. The conflict did not arise, and the new factories rose resplendent.

7

Opening the Market

From the first, Singer and Clark played quite different roles in the firm, in accordance with their different characters. Singer, as might be expected, was the partner more in view. The firm and the product were personified in the public eye by the inventor himself, who gave the Singer machine a selling image his rivals lacked. Grover and Baker, Wheeler and Wilson were mere names; Howe was known mainly for his litigiousness. The only begetter of the Singer machine, however, was by no means this kind of shadowy figure.

The *Atlas* in 1853 set the style for his image. "Isaac M. Singer is most emphatically a self-made, and, so far as it can truly be said of anyone, a self-educated man," it affirmed. This was an image well worth promoting, for it was very much the personality in vogue at the time. "This is a country of self-made men, than which nothing better could be said of any state of society," wrote Calvin Colton in 1844, a sentiment shared by a great many people. It was fashionable so far as the public was concerned (though not necessarily, as we shall see, so far as the fashionable were concerned) to be a self-made man. People wanted to read about these things; Harper and Brothers published a book in 1858 entitled *Self-Made Men*, consisting of a collection of essays outlining their careers. These tales reaffirmed people in their belief that there was real equality of opportunity in the New World, where ability rather than money or birth was the determining factor. Where such men happened to be inventors, they also personified the romance of this age of scientific progress. The diarist

George Templeton Strong spoke for a great many of his contemporaries when he wrote: "One thinks sometimes that one would like re-juvenescence, or a new birth. One would prefer, if he could, to annihilate his past and commence life, say in this A.D. 1860, and so enjoy larger acquaintance with this era of special development and material progress, watch the splendid march of science on earth, share the benefits of the steam engine and the electric telegraph . . ."

Singer was presented as, and in many ways was, the very personification of the spirit of this new era. The romantic tale of the invention of his machine was frequently repeated. Singer the great inventor was pictured continuously hard at work, and the fortunate public kept aware of its luck in being able to obtain the fruits of this creative mind fresh, so to speak, from the bench. "The leading characteristic of Mr. Singer's mind as an inventor is precision in the application of mechanical means," said *I. M. Singer & Co.'s Gazette*, distributed free to all customers and which first appeared in 1855. "He makes no mistakes—tries no abortive experiments; and although he has been able to improve greatly upon his own original ideas, yet his inventions have always been successful from the start, and have never been improved by any other person. He is now in the full maturity of his powers, and from the results of past efforts, much may be expected of him hereafter, for the benefit of his country and the world." One may imagine that George Zieber, who had now returned to Singer's as editor of the *Gazette*, had mixed feelings when penning these paeans of praise. (Never one to neglect an opportunity, Singer advertised in this same issue the sale of the patent rights for his defunct carving machine. "The Inventor being so fully occupied with the business of manufacturing Sewing Machines, that it is impossible to attend to the manufacturing and introducing of Carving Machines, is the reason why these Patents offered to the public," he explained.) In August 1855 the *Scientific American* took up the same adulatory cry. "Mr. I. M. Singer has become a Nestor in the discovery of Sewing Machine Improvements," it said.

Hardly a week goes by without the issue of one or more Patents for his inventions. His Sewing Machines have been much improved within the last year, until now they are in the highest degree perfect. Himself and Mr. Clark have already made large fortunes from the sale of their machines, and their business is rapidly increasing. We are glad of it. No one man has done so much towards the introduction of these great new sewing machines as Isaac M. Singer. He ought to be well rewarded.

The personal touch, and also the picture of confidence and success, was further enhanced by a "Grand Invitation Ball to be given by I. M. Singer & Co. on the evening of February 13th, 1856." Here, at no expense to themselves, patrons and business associates could meet each other and the great inventor under the most agreeable conditions. "We are not aware that any private mercantile or manufacturing Firm, has ever given a free public ball," remarked the *Gazette*.

We are accustomed, however, to doing original things, and believe that the precedent we intend to initiate is a good one and worthy to be followed by any one who can afford it. The arrangements have already been fully made with Mr. Miller, proprietor of the New York City Assembly Rooms, the most elegant dancing hall in town . . . Mr. Miller is highly pleased with the originality of the idea of a Private Firm giving an extensive ball, and paying every expense (as we intend to do in this case), and promises to use every exertion to make it the most recherché affair of the season. Cards of invitation will be sent in due time, to all persons who own Singer's Sewing Machines—to all who are employed in the manufacture of them, and to all who are engaged in operating them. These, together with the members of the newspaper press with whom we advertise, it is anticipated, will make a large and brilliant assemblage.

The ball was a great success. Three thousand people attended it, despite the bitter weather, and were able to enjoy refreshments, the music of Harvey Dodsworth's Band, and the sight of "five beautiful and talented young ladies busily engaged in sewing on the newest types of Singer machines."

However, Singer's at this time was very far from being the established money-making success all the publicity implied. While Singer happily blew air into the public bubble, Edward Clark, as might be expected, retained a more realistic view of affairs. For the truth was that I. M. Singer & Co., despite all the publicity, was not doing very well. Sales were not increasing to any substantial extent. In 1853, 810 Singer machines were sold; in 1854, 879; in 1855, 883. Meanwhile, it had been agreed that the settlement of $15,000 to Elias Howe should be paid at $5,000 a year plus interest. In addition to this impost, the winter of 1855 saw the start of one of those financial panics which swept the United States every twenty years or so during the nineteenth century. The last had been in 1837 and had been disastrous. "Two hundred and sixty houses have already failed," wrote the novelist Captain Marryat, who chanced to visit New York just then, "and no one knows where it is to end. Suspicion, fear and misfortune have taken possession of the city. Had I not been aware of the cause, I should have imagined that the plague was raging, and I had the description of Defoe before me." In the eighteen years since then trade had picked up and the markets had once more worked themselves into a frenzy; now the bubble was again about to burst. In January 1855 Edward Clark wrote to his father, Nathan: "During these hard times Mr. Singer goes on inventing. He has just made an important improvement . . . This is our consolation in regard to the sewing machine business. It relates to articles of practical use, not of mere pleasure or ornament. People will buy sewing machines when they will do without pianos." But they were not buying enough. On 2 April Clark wrote to William Proctor, then engaged on Singer business in Paris: "The winter has been an exceedingly hard one to us and everybody except moneylenders." "The state of trade has been so disastrous that Mr. Howe has just about swept away all the profits," he told a creditor the next day.

Once again the firm was in danger of closure and had to scratch for funds. George Ross McKenzie, a Scots immigrant, started with Singer's as a packing-case maker at $11.50 a week. "How's business, Mr. Singer?" he asked one day. "Terrible," came the reply. "We're about to be sold out." "For what?" "For lack of five thousand dollars." Once again Isaac was in luck. McKenzie disappeared for an hour and returned with five thousand dollars in cash, half of it his personal savings, the other half loaned to him by a bank which trusted him. A few weeks later, as the firm's capital was still tied up in stock and bills

Oswego in 1839. Courtesy New York Historical Society.

Rochester in 1812. Frontispiece to *Sketches of Rochester* by H. O'Reilly, Rochester, 1838. Courtesy The British Library.

Construction in progress on the Lockport and Illinois Canal. Note the hoist worked by horsepower on the same principle as Singer's rock-drill. Courtesy New York Historical Society.

It works! Zieber holds the lamp while Singer sews the first successful stitches ▷ on his sewing machine. Diorama reconstruction by the Smithsonian Institution, Washington, D.C.

Broadway, showing the section where I. M. Singer & Co. established their first offices, 1854. Courtesy New York Historical Society.

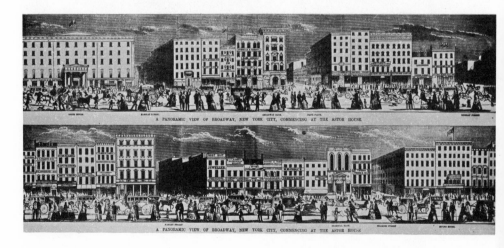

Right. The harbinger of civilization—Singer canvasser in Madagascar in the late nineteenth century.

Below

Left. Singer's first sewing machine, 1850. This model is on display at the Science Museum, London. Crown Copyright.

Center left. "Singer the Inventor." From the *Atlas*, 20 March 1853.

Center right. Portrait of Edward Clark. From Singer Manufacturing Company booklet issued on the occasion of the 100th anniversary of the Singer sewing machine.

Far right. Isabella Singer.

The Wigwam. From *The Architect*, 27 June 1874.

The Arena at The Wigwam. From *The Architect*, 27 June 1874.

THE SEWING MACHINE.

From *Harper's Weekly*, 10 March 1866.

Below left. From *Frank Leslie's Illustrated Weekly*, 17 April 1858.

Below right. Woman with Singer sewing machine, 1851. The packing case is used as a table and treadle stand.

The imprudent Agent of the Sewing Machine Company, having called upon Mr. Sanguine for a " Certificate," gets more than he came for, and goes off with flying colors—about the eyes and face. The man who has bought the machine for old iron rather enjoys the fun.

In the garden at Little Oldway. *Left*, Isabella and Singer. Children, *left to right*, Winnaretta, Franklin, Paris or Washington. Seated, *right foreground*, Alice Eastwood "Merritt" Singer, Caroline Virginia Singer.

Over page

Isabella Singer with Winnaretta ▷ in the garden at The Wigwam.

Singer in one of his party coats. ▷▷

△
Court scene during the contest over Singer's will. Mary Ann, the plaintiff, sits beside Surrogate Coffin. From *The Daily Graphic*, New York, 29 December 1875.

◁ Singer at The Wigwam, with Inslee Hopper, *right*, and George Ross McKenzie, first and third Presidents of I. M. Singer & Co.

Alice Eastwood "Merritt" Sing- ▷ er, alias Agnes Leonard the actress.

The wedding of Belle-Blanche. From *left* to *right*: Comtesse de Sadelys; Comtesse de Lowenthal; Duchesse Decazes; Elie Duc Decazes; Belle-Blanche; Duchesse de Camposelice (Isabella); Princesse Scey-Montbéliard (Winnaretta); Queen Isabella of Spain. From *L'Ar et la Mode*, 4 May 1888.

◁ Winnaretta dressed for a ball.

Winnaretta and friends. *Top row*: Prince Edmond de Polignac; Princesse de Brancovan; Marcel Proust; Prince Constantin de Brancovan; unknown woman; Leon Delafosse (supposed model for Charlie Morel in *A La Recherche du Temps Perdu*). *Second row*: Marquise de Monteynard; Winnaretta; Comtesse Anna de Noailles, née Brancovan (poet). *Front row*: Princesse de Caraman-Chimay, née Brancovan; Abel Hermant (novelist).

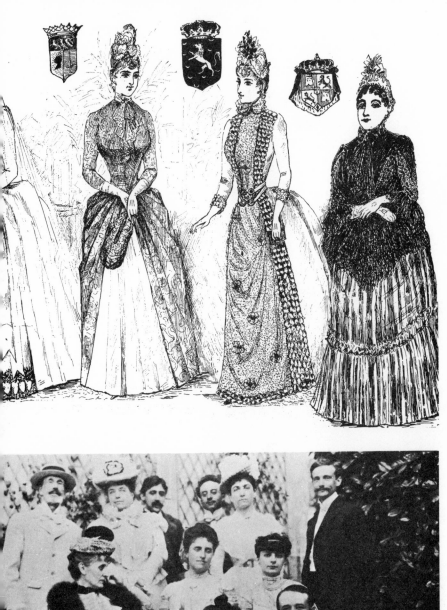

Above

Paris Singer and his wife Lillie.

Right

Isadora Duncan with Patrick, her son by Paris Singer.

were due to be paid, McKenzie agreed to take only three days' pay a week, and to credit the company for the remainder until things got better; he also managed to persuade most of the other employees to do the same. McKenzie received his just deserts: he ended up as the company's third president.

Nathan Clark also advanced the firm some money. By October 1855 Edward Clark could write to his father: "I am happy to say that business continues to be very flourishing, and I believe the time has arrived when no further borrowing will be necessary. I feel very grateful to you." That Christmas, I. M. Singer & Co. gave presents of dressing gowns to Nathan Clark and also to Ambrose Jordan and Charles Keller, the firm's lawyers, as "tokens of personal esteem" from the partners for "many kinds of favors of a substantial kind."

By the time the great ball was given, bankruptcy had been avoided and the firm was more or less on an even keel. But it was very far from making the huge profits attributed to it from outside and which had been so confidently predicted for it by both partners for years. In 1851, when Singer wanted to employ an experienced machinist, the man had asked, "Will the job last?" "Why man, it will be but a few years before I have a thousand men at work building sewing machines," came the confident reply. Similarly, in 1853 Clark had felt able to announce that "the success of our sewing machine is no longer problematical, it is a great fact." Neither of them now felt so confident.

Something had to be done to increase sales and profitability, and it was Edward Clark, the unpublicized partner, who did it. Although he had had no previous business experience, Clark now revealed a remarkable flair in his new field. The *Gazette* could justly claim that "we are accustomed . . . to doing original things." The originality— apart from the actual work of invention, a type of originality which had already been shown in the history of this very machine to be quite unconnected, in itself, with subsequent commercial success—was all Clark's.

★ ★ ★

Most of the early Singer machines had been heavy industrial models, designed primarily for use in tailors' shops and factories. Then, in

1856 the first family machine, designed for use in the home, was produced. It was called the "turtle back." It was rather smaller and lighter than previous machines and came in a wooden cabinet which could double as a table on which the machine could be placed for work. There were, however, considerable difficulties to be overcome before this new market, potentially so great, could be fully exploited. The difficulties were of two kinds: psychological and financial.

The financial barrier was perhaps the simplest to perceive. The cost of a machine was $125, while the average family income at this time was no more than $500 a year. Clearly, such machines must be out of the range of the very people for whom they were designed.

Such a situation was new to both manufacturers and householders. Previously, almost all goods needed for running the house could either be manufactured at home (such as household linens and various wooden implements and platters and pieces of furniture) or bought cheaply from shops or peddlers (such as pots and pans). Virtually the only essential household tool which was both expensive and could be bought only from the manufacturer and not reproduced at home was the gun—which was, as we have seen, the sewing machine's precursor in the field of mass production. Now came this new machine, an item clearly highly desirable now that it was available. It was, however, unlike the gun in one vital respect. There was no substitute for a gun; without it many households would have had neither food nor protection. But the sewing machine's work had been satisfactorily executed for centuries at a cost only of time and effort— and women's time and effort at that.

Something of the same problem had faced Cyrus McCormick when he started selling his reapers. His solution, when the farmer could not afford the full price for one, was to take a down payment plus freight costs, the remainder to be paid, plus six percent interest, the following December first, after the harvest had been reaped and sold. But even split in two, the price of a sewing machine would be an intolerable financial burden for most families. What was more, the sewing machine was not an investment which might add visibly to the family's future prosperity, as the reaper was to many farming families. The sewing machine simply saved time, and time was not money except when it was a question of getting the harvest in.

One possible solution to the problem was suggested by Louis Godey, the influential editor of *Godey's Lady's Book* and a strong supporter of the sewing machine: "One of our fair readers has been distressed because she cannot pay $125 for a sewing machine and since she has a family of ten children, a sewing machine is important. Many of you perhaps have been pondering the same problem, so we suggest that ten families in each country village unite to buy a sewing machine. Each could use it in an agreed period, then pass it on to the next family." It was not a bad idea, but it was enough to make any self-respecting businessman shudder. The problem was how to sell ten machines to the ten families for more than a tenth of the price.

Clark solved the problem by taking McCormick's idea several stages further. "Why not rent a sewing machine to the housewife and apply the rental fee to the purchase price of the machine?" demanded *I. M. Singer & Co.'s Gazette* in 1856. "Her husband cannot accuse her of running him into debt since he is merely hiring or renting the machine and under no obligation to buy. Yet at the end of the period of the lease, he will own a sewing machine for the money." The terms of purchase were easy: five dollars down and the rest to be paid, with interest, in monthly installments of three to five dollars. If customers could make a larger initial payment, they were encouraged to do so.

The scheme caught on at once and sales rose dramatically from 883 in 1855 to 2,564 in 1856. "We now make and sell at least four times as many machines weekly as we did one year ago," noted the *Gazette* on 1 October 1856. "Our workshops . . . now comprise four rooms instead of two." The notion of hire/purchase sales, as they were called, was quickly imitated by all Singer's competitors and has since become the standard selling practice for all domestic gadgets.

Although hire/purchase quickly became very popular, some viewed it with mixed feelings. The *Scientific American* commented: "A psychological fact, possibly new, which has come to light in this sewing machine business, is that a woman would rather pay one hundred dollars for a machine in monthly installments of five dollars than fifty dollars outright, although able to do so." Horror stories about machines all but paid for and then repossessed began to appear in the newspapers.

> There are hundreds of cases where poor women, widows perhaps, with a lot of children, are induced to buy sewing machines on this installment plan—five dollars a month, probably, for a forty-dollar machine [ran one typical such story: prices of machines had been halved by 1859 as a result of increased sales and modern manufacturing methods]. Perhaps they will manage to pay thirty dollars of this money and then get behind. I have known cases where women have been back in a small unpaid balance not more than ten dollars, for which they have been not only arrested and put in prison, but the machine has actually been taken from them and all the money paid on it kept as rent for the time they have used it.

Harper's Weekly and *Frank Leslie's Illustrated Weekly* ran cartoons on the subject. For his part Edward Clark, too, was not entirely happy about some of the results of his new scheme, which also carried some risks for the supplier. "It appears that $995.50 worth of machines were sold, of which you receive $76 in cash," he wrote the firm's Milwaukee agent. "This is too small a proportion . . . We had rather do a safe business in these times though small, than a larger one taking great risks. Try to make all sure and leave the doubtful customers to your competitors . . ."

The next year, 1857, the new Mott Street factory was opened in New York. It had the capacity to produce 300 machines a week, and once more something would have to be done to boost sales if such an investment were to be justified. Again, Edward Clark came up with a bright idea. The *Gazette* announced a "LIBERAL PLAN OF EXCHANGING SINGER'S NEW AND LATEST IMPROVED SEWING MACHINES FOR OLD OR UNIMPROVED SEWING MACHINES OF EVERY KIND. The time has now arrived when we are compelled to say, frankly, that it is impossible to add improvements to the old Machines of our manufacture so as to make them equal, or anything like equal, to the new ones." Readers had perhaps not yet learned, as have today's consumers, automatically to distrust any sentence from an advertiser containing the word "frankly." In fact the new machines differed very little from the earlier ones: they were merely newer, with a few attachments added. "THE PRICE WE PROPOSE TO ALLOW FOR OLD MACHINES, IN EXCHANGE FOR NEW ONES, IS FIFTY DOLLARS EACH," the announcement went on.

> We do not, however, restrict our offer to make such exchanges, to old Machines of our own manufacture. We will exchange on the same terms, for old Sewing Machines of any and every kind in use. Thousands of industrious and worthy persons of small means have had inferior or wholly worthless Sewing Machines palmed off upon them. The loss to these purchasers has been severe, but an infinitely greater damage has resulted to the public, in retarding the introduction into general use, of one of the most important labor-saving Machines of the present century. Sewing Machines of the Lerow and Blodgett, the Wilson, the Avery, and the Grover and Baker patents are lying idle, and useless, about the country, in great numbers. They were too imperfect in contrivance and workmanship ever to be used with success. These worthless Machines now stand directly in the way of the sale of good ones. Their existence causes great pecuniary loss to us. Probably all the Sewing Machines sold in the United States, in any year hitherto, has not exceeded ten thousand. Were the worthless Machines out of the way, we believe the yearly sale would exceed fifty thousand. We, therefore, have an extensive and direct interest in having all bad Sewing Machines finally withdrawn from the market, and our most improved ones substituted in their place. If, in six months, we can exchange our thousand new Machines for old ones, we have little doubt that our sales of Machines will be increased five thousand before the end of one year.

Meanwhile, Singer's was taking no risks of second-hand machines being resold cheaply on the market. "The old machine will be brought to our office in New York, and there be immediately broken up and destroyed," they added artlessly. Sales in 1857 rose by nearly half from 2,564 to 3,630.

By 1857 I. M. Singer & Co., despite the recession, was a prosperous and going concern. The firm's bookkeeper, Henry Milford, presenting the statement of accounts for 1856, could report that the year "has been a most successful and prosperous one in every respect. It affords me great personal gratification in being enabled to present to you, so favorable an account; and I would add, that such results must necessarily follow, where the *personal attention of the partners* is so wholly devoted to the business; such untiring exertion, perseverance and energy *must* bring its reward."

★   ★   ★

The new selling methods were thus obviously paying off. On the other hand, an annual sale of 3,630 machines, although a considerable improvement on previous sales figures, could still hardly be described as supplying a mass market. This could not happen until the industry was tooled up for mass production, as previously described. There was also a psychological barrier having to do with the fact that the person who would be using the family machine was, of course, the woman of the family. The poorest women were, traditionally, those who took in sewing and, once it was recognized that far from putting them out of business the sewing machine might be a boon to them, they were regarded as a natural market for it. The question now was its acceptance by the middle classes.

The concept of domestic labor-saving machinery was an entirely new one, and neither the lack of it nor the necessity for it was readily perceived. When such machinery was introduced in the factory or on the farm it was generally welcomed, especially in labor-hungry America. In the home its welcome was altogether less certain.

There were various reasons for this, the crudest one being that in most cases it was the man who earned the money and would have to pay for any such machine, while its sole beneficiary would be his wife. If a man's mother had run the home without a sewing machine, why could not his wife do the same? One of the functions of a wife was to make and mend clothes—beautifully and by hand. Why should he buy an expensive machine to do the same job altogether more impersonally? In 1859 *Frank Leslie's Illustrated Weekly* ran a cartoon sequence in which "Mr. Plumley, tired of stitching up his own coats, concludes to get himself a sewing machine. He accordingly gets one that will not get out of order, and will cost nothing to keep in repair [the picture shows him with a coy young woman]. After a few months of marriage, Mr. Plumley finds to his great dismay, that said machine does cost something . . . [the coy young woman having become, in the meantime, a nagging wife]." One feels that this attitude was one shared, if not openly avowed, by most men.

It is likely that many women too did not realize quite what an effect labor-saving domestic machinery might have on their lives until the *fait* was more or less *accompli*. In Europe and much of America, domestic help was available to all but the very poorest. Where, as in isolated homesteads or villages, such help was not available, hard, hard work was accepted as the necessary adjunct of a virtuous life. Roger Burlingame points out that "the women do not seem to have been aware of the terrible burdens of their lives until they were relieved of them. They simply accepted them as a part of their fate. 'Rest' was something which belonged in the next world; the word appears with great frequency in the Calvinist hymns." Now it was proposed that large sums of money that was scarce already be spent on a machine, possibly complicated to maintain and repair, whose only visible benefit was that it saved the housewife's time—to do what? The devil, as we know, finds work . . . *Harper's Weekly* ran a cartoon showing a salesman demonstrating a sewing machine to a group of ladies with the caption: 'A most Wonderful Invention, indeed, Mum, and it really Executes the Work so Efficiently and Quickly that, 'pon my Word, I think there's nothing left for the ladies to do now but to *Improve their Intellects!*" The absurdity of such an idea was regarded as self-evident.

To find something absurd is quite often a way of warding off its more threatening aspects. There can be little doubt that many men may have found the idea of women being given time to think not merely absurd but worrying. If a woman acquired a sewing machine, bought perhaps by herself out of the housekeeping money on the installment system and intended solely for the improvement of her own life, this was merely one more indication, and certainly a reinforcement, of a disturbing tendency which more and more women were showing at this time: namely, they wanted to take control of their lives into their own hands and out of the hands of their fathers, husbands and brothers, however well-meaning. The first national women's rights convention in the United States was organized by Elizabeth Cady Stanton in Seneca Falls, New York, in 1848. Mrs.

Stanton's Declaration of Women's Rights was modeled on the Declaration of Independence. In England, the long struggle for universal—as opposed to universal male—suffrage would soon begin. Freedom from household chores would of course mean that women would have more time to think about and participate in such movements.

Attacks on the new machine therefore often took the form, so universally applicable, of trying to prove that there were certain things for which women as a sex were fundamentally unsuited. Women, for example, could not and should not try to work complex machinery. It would be beyond them and it would end in tears. Newspapers printed letters reputed to have been received from ladies who had been quite unable to master the intricacies of the machine, such as this one from a California housewife who described how her dream of owning a sewing machine was at last rewarded:

> I always have, or *had*, lots of sewing to do, and although Mr. Adams is not poor, he soon would be so if I put it out, or even hired a seamstress in the house. And children do require so many changes of clothes in order to be kept clean and tidy! Last spring, my husband said if I could manage to leave home for a week, he would take me down to San Francisco to buy some drygoods and a sewing machine. He had always objected to me having one—not that he wanted me to be forever stitching with my hands, but because he thought I could not learn it readily, and that it would get out of order, which would be a catastrophe in such an out-of-the-way place as Coloma.

Despite these misgivings the machine was bought, and carried home in triumph.

> The next day my precious machine was unpacked, and following the printed directions, I succeeded in fixing the cotton and threading the needle, which was already properly set. It seemed so perfectly easy to work, as I had seen the man at the store operating upon it, that I spurned the idea of trying on a rag, and confidently put under the cloth-presser a leg of a pair of drawers for one of the children, that I had just cut out and basted. The whole of the ten young ones were standing around me, with my husband and the girl, and, looking upon them with rather a triumphant look, I said:

"Now you will see what a saving of labor this will prove," and put my foot upon the treadle and started off. The wheel made a half revolution forward, and then came back with great facility; my work moved the wrong way; the cotton became mixed up in a lamentable manner, and when I endeavored to pull the work into place, crack! went the needle.

Nevertheless the lady persisted for several days.

Early on Monday morning I was again at work, and this time I had the happiness of actually making stitches. To be sure they were in all directions, of various lengths, as I had pulled or held back the work, and their loose appearance was not very satisfactory to a neat seamstress, as I profess to be. Still, there was no getting over the fact that they were actual stitches . . . While I was blundering about, Johnny, my next to youngest boy, put his finger under the needle and his foot on the treadle, and sewed his dear finger in a shocking manner. In spite of this I was determined not to give it up . . .

Such a machine was clearly not a domestic boon.

Another much-used line was that, although ladies might learn to work a sewing machine quite successfully, this was not something that a respectable husband would really like his wife to do. "My wife —*my* wife—my *wife*—MY WIFE—listen to that—MY WIFE!!!—my wife wanted a sewing machine," spluttered Q. K. Philander Doesticks, the *New York Weekly Mercury's* humorist.

Of course, if she had asked for a small locomotive-engine, and a circular railroad round the parlor, she should have had them, or rather *would* have had them, and so I bought that sewing machine; and what is worse, I *paid* for it. If I hadn't paid for it, I could make them take it back; but as it is, I cherish no such delusive hope, and can only sit down here disconsolately and warn young husbands not to go and do likewise . . . I expect, every minute, that she will make a descent on my aquarium, and sew my fishes' tails together, or fasten the lobsters' legs to the breast-fins of the eels. How can I stop her? I appeal to husbands.

Such propaganda could only be met by counter-propaganda; and the tone of the sewing machine companies' advertising shows that they were never in any doubt as to where their great potential market lay. There is no record that any of the early purveyors of sewing machines were in any way personal supporters of the feminist cause—Singer, had he ever thought that any of the women in his life harbored such thoughts, would doubtless have been both disbelieving and appalled —but their advertising from the first was directed toward women, who were encouraged to think of themselves as independently in control of their lives. Principle was one thing, but commercial necessity could always override it. The Singer booklet *The Story of the Sewing Machine* had emblazoned on its front cover the legend: "Singer the Universal Sewing Machine, Sold only by the Maker Directly to the Women of the Family."

The most effective way to counteract prejudice was of course to demonstrate its illogicality. Doesticks might assert one thing, but the reality of life could be proved to be quite another. From the beginning, women were employed to demonstrate and teach the operation of the machine, thus proving beyond a doubt that they were perfectly capable of controlling it and would remain sensible and unhysterical while doing so.

One of I. M. Singer & Co.'s earliest employees was Miss Augusta Eliza Brown, who joined the firm in 1852, learned to operate the machine in two weeks, and from then on spent her time teaching others and demonstrating the novelty. She was employed to work her machine at the great fair in New York's Castle Garden in October 1852. The poor girl did not find this altogether easy: "Sometimes the needle would become unthreaded, and the thread would twist itself around the part we call the spring of the wire, and it confused me a great deal, because I had not been accustomed to work on that machine as much as the other . . ." Nevertheless, Miss Brown coped and remained with the firm for years. In accordance with his habit of employing members of the family wherever convenient, Singer used his niece, Mary Mariah Singer, as another of his lady demonstrators. (She was the daughter of one of his brothers, Elijah Singer, who left his family to go out to California on account of lung trouble and supposedly died there during the Civil War.)

"We must not forget to call attention to the fact that this instrument is particularly calculated for female operatives," pointed out an advertisement for Singer's Perpendicular Action Sewing Machine in 1853. "They should never allow it to be monopolized by men." To prove the point, "we have got possession of a front window under our office at the moderate rent of one thousand dollars a year," wrote Clark to an agent in 1852, "and a nice little girl is operating a machine in it, to the great entertainment of the crowd."

Having proved that women could successfully use sewing machines, the next necessity was to show that respectable women did so. Unfortunately respectable women tended to be neither rich (when they wouldn't care about such things anyway) nor adventurous. Some specific appeal had therefore to be made to ladies of unimpeachable respectability. Once again, it was Clark who hit on the solution to the problem. He published an appeal

To Pastors of Churches and Ministers of the Gospel, of every Denomination: The great utility of Singer's Sewing Machines, and the ample pecuniary profit derived from their employment, are established facts. The record of their steady improvement and brilliant practical success, makes one of the brightest pages in the history of American inventions . . . With these general remarks, we proceed to make an offer in the way of business, which we wish to be distinctly understood, as we shall not, in practice, deviate from it in the slightest particular. We will sell to any minister of the Gospel in charge of a congregation of any denomination, one sewing machine of our manufacture, of the most improved kind, and of either size designated, at one half the regular cash price, as stated in the printed list in *I. M. Singer & Co.'s Gazette* . . . This offer is designed primarily to the benefit of pastors of congregations and their families; but where a minister happens to have no family, or is so situated that the use of a sewing machine in his own household is not necessary or convenient, then we will supply the one Machine on the same terms, for a Sewing Society connected with the church, if there be one, or for the use of any deserving member of the congregation whom the minister may designate . . . We do not care to disclaim the general desire to do good to others; but the offer above made, liberal as it certainly is, is founded upon ordinary business calculations [Clark was always disarmingly frank about his motivation in such matters]. We know our Sewing Machines to be a good thing, having already made a large amount of money by the manufacture and sale of them, while the purchasers have been richly remunerated. Whenever one of our Machines is put to use, and especially if it be in a prominent place where numbers of persons have an opportunity of seeing its operation, other sales are sure to be made in the same society or neighborhood. For this reason, it is a matter of importance to us to have one of our Sewing Machines employed within the circle of each religious society in the United States.

The avowedly commercial motivation of this offer in no way discouraged ministers' wives from flocking to take it up. Most of them were poor and many had large families and distressingly unworldly husbands. "I. M. Singer & Co.: I want one of your *Transverse Shuttle Sewing Machines* Price $50," began one typical letter.

My Husband was formerly an able and successful Minister in Ohio. A few years ago he commenced laboring with a small salery [*sic*] on a late Missionary field in the West he became embarrassed—lost his health and has fallen into such a state of mind as to be unfitted for his calling and no longer supplies the pressing wants of his dependent family. I am trying to meet this deficiency by sewing. Can you extend me the favor shown to Ministers families and sell me the above named machine at *half price*?

The reply to this plea is not recorded; one suspects that the request might have smacked too much of genuine charity and carried too little commercial advantage for Clark.

Meanwhile advertising copy was presenting the increased free time which the sewing machine allowed the housewife as a positive virtue rather than an alarming innovation. The key question, of course, was: free for what? "The great importance of the sewing machine," announced a Singer booklet,

is in its influence upon the home; in the countless hours it has added to women's leisure for rest and refinement; in the increase of time and opportunity for that early training of children, for lack of which so many pitiful wrecks are strewn along the shores of life;

in the numberless avenues it has opened for women's employment; and in the comforts it has brought within the reach of all, which could formerly be attained only by the wealthy few.

It would be a hard husband indeed who could object to *that*. A Grover and Baker pamphlet entitled "A Home Scene, or Mr. Aston's First Evening" put a more purely cultural emphasis upon the possible benefits to be derived from the machine.

Mr. A. feels his wife needs a change—a rest, and suggests they go to a concert being given at Niblo's that night. She: "I would enjoy the music very much, but am so much in arrears with my sewing, that I cannot afford the time . . . I find it difficult to get a seamstress to the house; besides, it is harder to give directions and make one understand what my wishes are, than to perform those duties myself."

Mr. A. notices an advertisement for Grover and Baker:

"Is there a husband, father or brother in the United States, who will permit the drudgery of hand-sewing in his family, when a Grover & Baker Machine will do it better, more expeditiously, and *cheaper* than it can possibly be done by hand?" Throwing aside the paper he started to his feet and said, "I cannot withstand that appeal! I must go and see these Machines! I must have one! Mary, you shall have your evenings, aye, and your afternoons, too, for relaxation and mental culture! I must have been asleep not to have seen through all this before!"

Most of these advertisements were obviously designed in the first instance to appeal to the men who would be paying out the actual money at their wife's behest. One Singer advertisement, however, made a frank appeal to the spirit of female independence: "The great popularity of the machines may readily be understood when the fact is known that any good female operator can earn with them ONE THOUSAND DOLLARS A YEAR," it claimed.

The point has been made that it was not the advent of the sewing machine in the home, but of cheap and good ready-made clothes in the shops, which finally liberated women from the drudgery of the needle. But a start must be made somewhere and, by the mid-1860s, the "Song of the Shirt" was becoming more a historical curiosity than a telling comment on current social conditions. Sewing parties of ladies learning the use of the machine, first mooted by Louis Godey during the Civil War, became all the rage. "I have made a set of pinafores for Little Emily's birthday," noted one enthusiast in her diary in 1866. "I find I am neglecting my Bible reading for these occupations and must be more watchful." Her minister would have been horrified, but Edward Clark, despite his piety, would have been delighted.

★ ★ ★

The sewing machine, then, was becoming increasingly more widely accepted and sales were rising. Of the various makes, Singer's may well, even in the 1850s, have been the best known by name, since this was the firm which introduced the various aforementioned selling innovations and since Singer himself tended to keep it in the headlines. But it was not, in the early days, as commercially successful as the Wheeler and Wilson Manufacturing Company's machine. Until 1856 the two firms sold roughly equal numbers of machines. But in 1857 Wheeler and Wilson sold a thousand more than I. M. Singer & Co.; in 1858 they sold 7,978 to Singer's 3,594; in 1859, 21,306 to Singer's 10,953; and so on in much the same proportions until 1867, when Singer's finally took the lead, never to be overtaken or even approached again. Edward Clark's policies were by then beginning to pay off and it was Singer, not Wheeler and Wilson, which became the household synonym for a sewing machine. How was the firm's long-lasting success achieved? It was obvious from the beginning that in a country the size of the United States no firm could hope to operate nationwide solely from Boston or New York. Indeed, in Singer and Phelps's very first advertisement, on 7 November 1850, they said that "an Agent (with whom exclusive arrangements will be made) is wanted in every city and town in the United States." In fact, however, the distribution of Singer machines in the early days was not usually undertaken by agents, whose establishment might entail the company in capital expense which it could ill afford, such as the opening, equipping and stocking of branch offices. There was no question that sales opportunities might be lost if the stock was not there—as a Mr. Shea found out when he canvassed Dayton, Ohio, in 1851: "The last two [machines] you sent arrived Saturday evening 18 October," he wrote.

> I sold the two fast and think I could have disposed of more if I had been in possession of them when I had caused the excitement at first. As it is I may sell more, one or two perhaps, but there is by no means so favorable a prospect as at first. I went down to Cincinnati on Saturday morning and returned same evening—could do nothing as I had no machines and more than half my time has been spent in vain as I have had to wait for ten days at a time for the machines which you had sent but which failed to arrive.

But the company in its early days was, as we have seen, very short of capital and simply could not afford to keep a large number of agents supplied with machines. They therefore, on a number of occasions, managed the problem of distribution not by establishing agencies but, as Blodgett had advised, by selling territorial rights to the patent.

Several advantages were supposed to be inherent in this system. One was that instead of paying an agent to set himself up in a district, the rights holders actually paid the company for the privilege. They also paid the company for the machines with which they were supplied, buying at a discount and later making a profit from sales at full price. However, the system had disadvantages. One was that not much capital was raised from the sale of rights, since at the time these were a highly speculative investment and not much could be got for them: there was not as yet any proof that the patent would turn out to be profitable. Another disadvantage was that once the rights had been sold the company no longer had any real control over distribution in that part of the country. They could not put in one of their own agents because this would undercut the rights holder; on the other hand, if he was lazy or inefficient they would be faced with the galling prospect of their rivals picking up all the sales. Where the rights holder was energetic, he might make very large profits for himself. The rights holder for Pennsylvania and Wisconsin at one point made estimated profits of $8,000 to $10,000 a year. But this did not mean that the company was making equally large profits on his activities, because the discount at which the machines were allowed to rights holders was so large. At first they got $125 machines for $60, later raised to $70 for improved machines. Such discounts, despite the savings on wages and offices, did not, if one bears in mind the costs of new tools and expanding workshops, even cover production costs. (The largest commission ever paid by the company to one of its own agents was 30 percent.) Despite its apparent short-term advantages, therefore, the system was clearly unsatisfactory. "In no instance have we sold any territorial rights that we did not afterwards regret it," wrote Clark; and in 1856 the company began a positive policy of buying back territorial rights where they could. Sales of Singer sewing machines were to be handled henceforth entirely by Singer's own agents.

The setting up of an agent was not quite the straightforward affair it might have appeared. "Singer machines cannot be bought of irresponsible dealers," affirmed a company pamphlet in 1900; and the same was true—or Clark did his very best to make sure it was true—from the very beginning. Great thought was always given to the type of persons employed, and also to the setting in which they operated.

An idea of the surroundings considered appropriate to the sale of sewing machines may be gathered from the company's own description, in the *Gazette*, of the new New York office in Mott Street, opened in 1857:

There is no place of business in the city more beautiful and attractive than this. A palace of white marble, tasteful and elaborate in the architecture and decorated interiorly with all the splendor of modern art; it forms an appropriate temple for the exhibition and sale of the most useful invention of the present century. For a lease of fifteen years of this building we paid a premium of twenty thousand dollars, besides assuming an annual rent of about thirteen thousand five hundred dollars. The fitting-up of the premises has also cost us a very great amount; very extensive alterations and improvements being required to make them complete for our business. But although the aggregate increase of expenses, thus incurred, has been quite considerable, we are quite satisfied, after a brief trial, that the change was a most judicious one, and entirely remunerative in a pecuniary point of view. We invite all strangers who have occasion to visit New York, to call at our office. There are few places of amusement even, which will better fill a vacant hour and the machines are always exhibited with courtesy whether there be any intention to purchase a machine or not. In our Office no person is ever importuned to buy a machine, or in any way subjected to annoyance.

While agents were not expected to get themselves up so lavishly, they were encouraged to observe the tone set. Thus, in 1856 Clark wrote to one agent: "We entirely approve of . . . your determination to wait until you can open in good style. There are many persons who are governed mainly by appearances." In the same year a prospective agent in New Orleans was advised that an investment of $5,000 would be needed to set up an agency there in the appropriate style, "such a place as ladies would not hesitate to visit."

The agents themselves had also to be carefully selected. One desirable quality was that they should be well acquainted with the sewing machine and all its works, so as to be able to demonstrate it satisfactorily to new customers and repair machines in their territory when they went wrong. In short, the best agent was a competent mechanic; but, especially in the early days, such specialized mechanics were rare and could ill be spared from the workshop. Nevertheless, competent salesmen were so important that several such men were put on the road. Orson Phelps was one such, much against his will, it will be remembered. Another was John H. Lerow, one of the patentees of the Lerow and Blodgett machine. Tailors also made good agents, being, so to speak, a living recommendation of the product; both the early Ohio agents in Cleveland and Cincinnati were tailors.

Another quality looked for in agents was loyalty to the firm. Loyalty was particularly important in view of the cut-throat competition offered by firms such as Wheeler and Wilson and Grover and Baker. I. M. Singer & Co. had to be able to trust their agent to sell and recommend solely Singer machines, despite the fact that much more money might be made by an agent if he represented more than one company. To this end, some of the more important agencies were entrusted to friends and relatives. The agency at Richmond, Virginia, was given to Benjamin Trott, one of Singer's brothers-in-law. Edwin Dean, the actor who had been Singer's first theatrical employer in 1830, was walking down Broadway one day when he noticed a crowd standing outside a shop window. He joined the crowd, which turned out to be watching the Singer seamstress demonstrating her machine. Could this, Dean wondered, be the same Singer he had known all those years ago? He went in to find out and came out with the St. Louis agency. Another reliable agent was Charles Sponsler, Mary Ann's brother and another old theatrical comrade. He operated from Baltimore. Agents received a discount, usually twenty-eight percent, although the discount allowed them went down when prices were reduced by half in 1859. The agents mostly resisted, unsuccessfully, both the reduced price and the reduced discount. However, if they were loyal and hardworking, the company reciprocated by giving them every support and sticking by them through thick and also through thin. In fact, once an agent had been appointed, they were reluctant to replace him even if he was unsuccessful. "An application has been made to us for the agency of machines in your place," they wrote to one such in 1860. "We think a good business might be done in your section if properly looked after and will so far depart from established custom as to furnish you a stock of machines to be accounted for where sold, if you give us security, or satisfactory New York City reference. Could you not make it profitable, if thus furnished, to go through the section of country adjacent to you, promoting sales?"

The tradition of hard salesmanship was not new to the American housewife; the art of selling had been practiced in a considerable variety of forms by Yankee peddlers. One such old-established technique relied on "soft sawder and human natur' "; it was practiced by Sam Slick, the clock peddler, who would leave his last clock to be looked after by the housewife when he had to leave in a hurry.

Now, that clock is sold for forty dollars. It cost me just six dollars and fifty cents. Mrs. Flint will never let Mrs. Steel have the refusal; nor will the Deacon larn until I call for the clock . . . how hard it is to give it up. We can do without any article of luxury we never had, but . . . it is not in human natur' to surrender it voluntarily. Of fifteen thousand sold by myself and my partners in this Province, twelve thousand were left in this manner; and only ten clocks were ever returned . . . We trust to soft sawder to get them into the house, and to human natur' that they never come out of it.

As for the idea of "loss leading" via the leaders of the community, that had first been practiced by Jim "Jubilee" Fisk. Fisk made his first money peddling Paisley shawls in association with Volney Haskell, a traveling jeweler. One of the partners would go ahead to a small town, seek out a well-known and good-looking woman, and give her a Paisley shawl, suggesting she wear it to church the following Sunday. There would be a stir of envy in the church when she "sashayed" up the aisle; and the next week, when Haskell or Fisk appeared in town with a carriage load of Paisley shawls, he would find a ready market among the women of the community. (Fisk went on from these modest beginnings to finance the Erie Railroad.)

But although the Yankee peddler with his smart tricks and wooden nutmegs was an established part of the American tradition, the country had never seen a concerted attack on the domestic market to compare with that of the sewing machine salesmen, and it aroused considerable public comment and opposition. The insistence of these agents in their eagerness to sell a machine before their prospect was snatched away by a rival, in many cases regardless of whether the recipient could really afford a machine or was likely to keep up the payment on it, was the source of many stories in the newspapers. A typical one came from Baltimore in 1863:

Jacob H. Aull, a sewing machine company's agent, was tried in the criminal court yesterday for assault on Mary H. Beard, who had rented a machine by the month and was in arrears for some months. Aull went to the house and removed the machine, in doing which Mrs. Beard charged he struck her. Aull testifies that he seized the machine and carried it off, though Mrs. Beard held on to part of it. He denied striking her or using unnecessary force.

Judge Pinkney said he would have to acquit the defendant, the Court of Appeals having decided that the agent can remove the machine on default in payment of the installments contracted for. There was sufficient resistance on the part of Mrs. Beard to exempt the agent from the charge of assault, but he was extremely sorry that the law was such as to permit this kind of business to be carried on . . .

Another newspaper quoted a well-known lawyer as saying that "'If ever there was a class of people that needed legislating against it is these sewing machine agents and their employers.' The system carried on by these agents has long been a subject of complaint," continued the article.

Time and again there have been publications in the newspapers to the effect that such and such a family is being sold out for arrears on a machine, and although there have been plain statements in regard to the way in which these men carry on their business they have never amounted to anything . . . The agents are backed by their employers and nothing seems to intimidate them. Now and then a flagrant case comes to the ears of the public and there is a general burst of sympathy for the unhappy victim and a general burst of indignation against the agent and his confederates, but it soon blows over and is forgotten. To this fact more than any other perhaps is owing the boldness with which these men carry on their operations. They carry their cow-catcher contracts around in their pockets and if they can put a machine in a poor man's house and get his wife to put her name or mark to this written instrument their end is accomplished. The absence of the husband is their chief reliance in the majority of cases.

This was a point frequently made, it being felt by the writers of editorials that men would be immune to the machine seller's guile. Naturally a man would be less motivated to buy a machine than his wife, but the real point was that women were not fit to handle financial affairs. Thus public opinion (or at least those who wrote newspaper columns) disapproved both ways: it was felt that sewing machines were a very dear luxury to ask a husband to buy for his wife; on the other hand, women were by no means considered capable of buying them for themselves.

Nevertheless, and despite such difficulties, sales continued to rise and rise. The estimated number of machines in America in 1862 was 300,000, of which about 75,000 were in use in private families. The capital vested in real and personal estate in the sewing machine business in nine states during the year ending 1 June 1860 was £289,656 (the figures are given in pounds sterling since they are from an English journal, the rate of exchange being about five dollars to the pound). The cost of labor that year was £221,520, and 116,330 machines, valued at £1,167,780, were made; 2,194 hands were employed in making them. The speed with which this new industry had developed was regarded as something remarkable even in that age of rapid change and progress. The *Practical Mechanic's Journal* commented in 1858:

> So lately is it since we first heard of an actual working sewing machine, and to such perfection of work has the system now arrived, comparing the results achieved with the space within which they have been accomplished, that it seems as if the clue leading to the mysteries of mechanical stitching had long been hidden or unnoticed—but when laid hold of, the whole matter was at once laid bare, and turned to practical account, in supplying a leading want in the arts of human existence.

This general prosperity, although showing that sewing machine manufacturers as a whole were doing well, still does not explain why Singer finished up doing so much better than any of the rest of them. In fact the secret was not to be found in American sales figures, impressive though these might be, but was from a very early date attributable to foreign sales. I. M. Singer & Co. was possibly the first of what has since become one of the most familiar phenomena in Western economies: the American-based multinational corporation. As early as 1861 Singer & Co. was selling more sewing machines in Europe than in the United States, and by 1867 it had established its own factory there (in Clydebank, near Glasgow) controlled not by a European rights holder but by the parent company in New York.

Most of the American sewing machines were available abroad, usually through a general agent who handled more than one line.

I. M. Singer & Co. from the first took the prospect of foreign markets a great deal more seriously than this. Not only did they establish their own agents in important centers—the Paris agency was opened in 1855, the Glasgow★ branch office in 1856, the Brazilian agency in Rio de Janeiro in 1858 (this was opened by George Zieber, unable any longer to stand the sight of his erstwhile partners getting richer and richer in New York)—but they very quickly began manufacturing their machines abroad as well. In 1855 the French patent rights were sold to a M. Callebaut of Paris. "We are happy to know that you have become the purchaser of our French patent, though the price presently paid is very inadequate, compared with the real value of the invention," wrote Clark in the spring of that year. (The price paid was 10,000 francs cash, another 20,000 francs to be paid as soon as orders for 30 machines were received.) William F. Proctor, one of the firm's most competent workmen, and an old acquaintance of Singer's from the days when he was building his printing machine at A. B. Taylor's, was sent out to facilitate production and look after the company's interests in France.

Setting up a factory and sales organization so far away from New York was no easy matter, especially since, as we have seen, the necessary machinery was new and expensive and its quality and accuracy were of the first importance. All the machinery for the new factory was to be sent from America but, wrote Clark, "The steam arm planing machine we shall not send, but instead of it will forward a good hand planer, which we know to be worth twice as much as the steam arm for the work you will have to do—besides which, we can procure the hand planers and the steam arm can't be got, at least without much delay . . ." In July 1855 he wrote: "No reasonable effort shall be wanting on our part to get you fitted out with an establishment which shall be a model one, and a credit to American skill in the mechanic arts." Then came the punch line. "At the same time we must say that we wish M. Callebaut to forward funds as fast as possible. Our very

★ The office was set up in Scotland rather than England because William Thomas, the purchaser of Howe's patent in 1846 in London, was still enforcing it in England and suing infringers. He could not carry his suit to Scotland or Northern Ireland, however.

large expenses last year and small business, comparatively, makes it necessary to use all resources."

It was not long before I. M. Singer & Co. had very good reason to be grateful for their substantial foreign establishment. During the 1857 panic Singer's kept going very largely upon the remittances from their foreign agencies. On 21 October of that year, Clark wrote: "The present dullness in the business is not because our machines are less saleable but because industry is paralyzed. We have cut down our expenses to the lowest practicable point . . ." Clark's difficulty was that the company was engaged upon an expensive building program, namely the much-vaunted Mott Street factory being constructed by D. D. Badger and Company. On 26 October, Clark tried, but failed, to borrow $25,000 from the firm of William B. Astor in order to keep going. Nevertheless, on 31 October he was able to write to Badger: "We have received a remittance from our agency in Scotland . . . and you shall have the benefit of it. The enclosed check for $1,000 will at least show you how good our intentions are." "We did discharge the greater part of our hands," he wrote on 14 November to Robert E. Simpson, the Glasgow agent, "but now we have got 100 of them at work again, and are gradually increasing. Our manufactory keeps going up notwithstanding hard times. Nearly all other firms engaged in building have had to stop."

Four years later, with the outbreak of the Civil War, Singer's once more had occasion to be grateful for their foreign connections. The war naturally affected trade, as it affected every other aspect of American life. Firms and private individuals had no spare money and no inclination, in such uncertain times, to invest what they had in expensive machinery. Agents in the southern states were virtually immobilized and cut off from New York. "We think the best course is for you to retire to your old home," wrote Clark to Wheeless, the Mobile, Alabama, agent, "taking such stock as you have and there . . . make the best you can of it. We have all confidence that you will pay as promptly as you can . . . Perhaps the interior of Alabama and Mississippi may not suffer much unless taxation shall weigh you down. New Orleans and Mobile will very likely be prostrate." To the Nashville, Tennessee, agent Clark wrote: "Immediately upon receipt of this, we wish you to store what merchandise we have in Nashville in some safe place and in care of some respectable and responsible person, and *you and your wife by the most direct and cheapest route to New York*. We . . . wish you a pleasant and safe journey. While the war lasts it may be as well for all engaged in the arts of peace to accept the fact and cease to do business." Clark was an ardent supporter of the Union cause but many of the southern agents were equally ardent Confederates. While deploring their opinions, Clark did not allow this disagreement to come between them in a business sense. "For your personal welfare you have our best wishes and for the cause you favor all the success it merits. Perhaps your opinions may change, at least we shall hope so," he wrote Wheeless.

Where business could continue, it did not flourish. The greatest difficulty was in raising ready cash. Nor was Singer any help. Having been forced by personal problems to leave the country, he was then in England. "I expect to realize the proceeds of 7,750 bushels of corn which is on the way from Chicago and was bought because exchange could not be had at any reasonable prices. From that course, it is likely about $3,000 will be realized," wrote Clark to Singer in May 1861. A year later, business had still not improved. "Business has been quite dull for the last three weeks," wrote Charles Sponsler from Baltimore in April 1862. "The few government contracts given to this city have been filled, and the clothing men have done very little this spring." Everywhere the news was the same. "We are scudding along under just as close sails as we possibly can and we trust to come through all right," wrote Clark, in June 1862, to Singer, who was still abroad and blithely unconcerned with the crises at home (he was, as we shall see, concentrating on other matters).

Unless this war is brought to a speedy termination more than one half the mercantile houses in this city must close. Nobody is making money except those who have been fortunate enough to get some war contracts connected with the equipment of troops and very few are doing business enough to pay expenses . . . The movement of the rebels in removing their forces from Harper's Ferry and concentrating on Richmond indicates a great battle soon.

Clark reconciled the demands of political conviction with the difficulty of disposing of surplus stocks by donating 1,000 machines to the Union cause, to sew uniforms for Grant's army (to this end, Louis Godey was urging ladies to form sewing bees and take instruction in the use of the machine)—a move which provided much useful publicity when the war was finally won.

The rest of the world, however, was not suffering from a slump in trade. On the contrary, business was flourishing, especially in Britain, which was approaching the height of its free-trading prosperity. In most countries import tariffs were low or did not exist at all. Moreover, because of the war the value of the dollar was falling, so that remittances in foreign currency were more and more valuable. Before the war, Singer's foreign representation had not, as it happened, been entirely satisfactory. The Paris venture had turned out disappointingly and had ended with lawsuits. In 1861 Singer discharged the London agent and wrote to Clark, "My belief is that if we had never had anything to do with foreign countries and had attended more strictly to that of our own we should be much better off today." This was one of the biggest errors of judgment he could have made. We have already seen that remittances from abroad had been of great help to the company, and before the Civil War was over Singer's was sending abroad all the machines it could not sell at home: foreign trade was booming. The foundations for what was to become a huge international trade had been established. In 1863 a branch was opened at Hamburg in Germany, and in 1867 the firm established its factory in Glasgow. Thus a new trend was becoming apparent. Before this the New World had exported its raw materials and, almost alone among the world's trading nations, had erected high tariff barriers to protect its own manufacturers against the manufactured goods of Europe. But now Europe and the world would become a market for the fruits of the newly established American mass-production techniques.

From now on the foreign market was to become more and more important to Singer's. This was the answer to the question of why Singer's outstripped their rivals. It was not just the sewing machine, but the Singer sewing machine, which became a byword as the first and often the only aspect of Western civilization to reach the remotest backwaters of the world. "THE HERALD OF CIVILIZATION," ran one nineteenth-century Singer advertisement.

MISSIONARY WORK OF THE SINGER MANUFACTURING COM-PANY. At the close of the recent war, the King of Ou (Caroline Islands) came to pay homage to the Government at Manilla. As the best means of advancing and establishing a condition of things that would prevent all future outbreaks, the King was introduced to the "Great Civilizer," the Singer sewing machine, and we have here his photograph, seated at the Singer sewing machine, with his Secretary of State standing beside him. This is absolutely authen-tic. It is a half-toned plate made from the original photograph, which can be seen any day at the office of the SINGER MANU-FACTURING CO., 149 Broadway, New York City.

8

Mr. Mathews, Mr. Merritt and Mr. Singer

This examination of the commercial development of I. M. Singer & Co. has got somewhat ahead of the story. For while all these setbacks and successes were naturally of the greatest importance to Singer himself, in that they provided—or at times failed to provide—the dimes which it had, from the first, been his ambition to acquire, they did not constitute the ruling interest of his life, as they did, it would probably be correct to say, of Edward Clark's. An obsessive businessman would hardly have been still in bed when Zieber came to call on him that May morning in 1851, to quote but one instance. In the business, Singer got on with his inventing and improving and on the whole was happy to leave questions of policy to his partner while collecting a gratifying amount of publicity for himself. All this, however, left him plenty of time to pursue his personal life—which grew more and more complicated and must have consumed increasing amounts of energy as the decade wore on and the dimes rolled in.

When we last met the Singer family in 1851 they had just graduated from the overcrowded and rather squalid conditions of the Lower East Side which Gus Singer recollected from his boyhood, where Isaac had to tinker with his invention and conduct business in the room in which Mary Ann had just given birth. They moved to a house on East Fifth Street, near Fourth Avenue, and in 1852, as their prosperity continued to mount, they moved again to 374 Fourth Avenue, where they remained for some years.

Mary Ann and Isaac's family life at this time was a model of

Victorian respectability. They acquired the objects appropriate to the home of a prosperous businessman: a grand piano, expensive pictures, ornate furniture. They kept carriages on Fifth Avenue. In 1854 the family was visited by the usual Victorian tragedy of child mortality: the two youngest children, Charles Alexander and Julia Ann, died aged, respectively, four and two. A family plot was purchased at Greenwood Cemetery and there the children were buried. Their graves were marked by both a tombstone bearing their names and a monument, personally selected by their father: Mary Ann remembered accompanying him to select the stone.

Singer enjoyed displaying his new-found status—as who in his position would not? He had not been born to wealth and luxury, nor even to comfort. He had known prolonged periods of the grimmest poverty. He was a man acutely conscious of the impression he made on others; his flamboyant personality craved the applause which, after that brief moment of glory as Richard in Rochester more than twenty years before, had been so totally denied him. His failure as an actor—his chosen profession and his passion—must have been horrid to him, especially when he saw other men with neither his grand physique nor (he doubtless considered) his talent achieving the success which had eluded him, and for which his business success would never be a substitute. But in this attitude he was not typical of the American public. For them entrepreneurs and inventors, now at the beginning of their great careers in the public eye and not yet relegated as they would be in the twentieth century to the anonymity of the boardroom and the commercial laboratory, were glamorous figures. Were they not the embodiment of the American dream? In the early years the political idealists and social experimenters had had their day. The Thomas Jeffersons and Tom Paines, William Cobbetts and Robert Owens had been all very well in their time, but the political and social idealism they held out was not what lured the immigrants in their millions from Europe to America in the 1850s, nor what sent the forty-niners to California. What such people were fleeing was economic rather than political oppression (although it is difficult and perhaps misleading to separate these too definitely). What drew them on was the near certainty of economic independence and the possibility of, somehow, getting rich quick. In such a society actors were an irrelevance and a frivolity (as Frances Trollope had noted). It was people like Isaac M. Singer the inventor of the sewing machine whom everybody wanted to emulate.

The personal attention Singer received in his role as a "Great Inventor" was both good for the company and, naturally, personally gratifying. Always inclined to see a situation in dramatic terms and to act it out both on and off stage (it will be remembered that he took the whole business of the Sewing Machine War quite personally, convinced that all his competitors were pledged to deprive him alone of his just deserts and that they would be punished hereafter for their presumption), he now assumed the mannerisms appropriate to this new role. Toward his employees he adopted a paternalistic stance. The Fourth Avenue house was thrown open to them every New Year's Day when all were invited to visit the boss at home and enjoy being welcomed and waited upon by his wife and daughter Voulettie, now in her teens. He had always kept a generous house even when he was poor— Zieber had testified to this: now he could afford to entertain in style.

Such performances evoked varying reactions from those who had known him in his earlier, less prosperous incarnation. Mary Ann can have felt nothing but relief and possibly a sense of gratification: the seemingly rash decision of nearly twenty years earlier, when she had thrown in her lot with the irresistible young actor and taken him on for better or worse but without benefit of clergy, was at last surely justified. As for the latter detail, it must have assumed increasingly less importance. For all practical purposes and in the eyes of the world, Mary Ann now *was* Mrs. Singer. That was how Isaac always addressed her before strangers; that was how he introduced her to his partner and employees, and that was how they knew her. When she called in at the office, as she sometimes did, to pick up some money from the cashier, she was received and obliged as the boss's wife. At other times, when she went shopping in emporia such as Stewart's or Lord and Taylor's, she would not pay in cash at all but simply sign for the goods as Mr. Singer's wife, leaving him to settle the bills. She was, in short, a thoroughly respectable woman as far as society was concerned; and who was to say her nay?

Well, Catharine, William and Lillian for a start, but they did not seem disposed to do so. Singer had by no means abandoned or disowned his first family: we have seen that William was acting as one of the firm's agents in 1851 and was known by all as Singer's son. It is impossible to know quite how this was explained away, or even if it was felt worthy of any explanation—though, human nature being what it is, there must have been some speculation. Did many people simply produce grown sons out of a previous life in those days? As in earlier phases of Singer's life, when he picked up Mary Ann and persuaded her to come and live with him and, equally, as when his mother walked out on her husband and family, there seem to have been certain assumptions of freedom of choice and action on the part of women as well as men which were quite different from anything obtaining in Europe at that time.

One cannot merely assert that Singer was the feckless philandering son of feckless parents, although many would have liked to label him as such: this implies, after all, that the other parties to such philanderings had themselves no freedom of choice or action. Perhaps the sheer size and uncertainty of communications within the United States had something to do with it, allowing the possibility of a different life in every state, just as European sailors were reputed to keep a different wife in every port. However, whereas the advantage of this is the possibility of keeping various wives, lives and families from overlapping in an embarrassing manner, Singer always seemed blithely unconcerned about such delicacies—and so, more surprisingly, did the women involved. Perhaps they were simply carried away by the force of his total disregard for awkward proprieties. There is no evidence that Mary Ann, for example, was in any way what could be called common, or that her family and background was not respectable. She kept in touch with her family and they did not disown her—although there is no clear indication of whether they actually suspected that she was not married. Everyone who met her remarked on her natural dignity, kindness and good manners. It might be argued that force of unforeseen circumstances had landed her in her present situation; but it is inconceivable that any European girl of good or even mildly respectable family would have been allowed to travel alone from her hometown to another two hundred miles away at the request of, and in order to meet up with, a young man to whom she was not married. It was of course this freedom of action and decision, already noted, which gave American girls such an edge over their European counterparts in late nineteenth- and early twentieth-century European society, thus providing Henry James with so many of his plots. The point to note in this connection is that this freedom in no way denoted a lax moral tone. On the contrary, it was a reflection of the real force of the Puritan ethic in American society. It implied that American girls—and American men—were to be trusted and needed no chaperoning. It was simply Mary Ann's bad luck that in this as in every other way Isaac was a stranger to the Puritan tradition.

At any rate, it seemed to have all turned out for the best; and even though Catharine was now living in Long Island, not far away, she was evidently keeping quiet, having long since found herself a much more satisfactory partner by the name of Stephen P. Kent and presumably concluded that the less said the better. For Mary Ann, then, the gamble had paid off. Twenty years and eight children after their first meeting in Baltimore (two more, Julia Ann and Caroline Virginia, were born in 1856 and 1857 to replace the dead Charles and Julia) she must have felt as assured of her position as, given Singer's basically exaggerated and unrealistic temperament, it was possible for a woman to feel. She could and did pass around the refreshments with a good grace.

For George Zieber, on the other hand, the sight of his erstwhile partner was particularly galling. When he first left I. M. Singer & Co. Zieber went to England to see what he could make of the sewing machine business there, but he did not do very well and returned to New York where he was given the editorship of *I. M. Singer & Co.'s Gazette*. "I applied myself diligently to the work of enlarging and extending the business," he wrote, "thinking there might possibly be lurking in Singer's heart an intention to do what was right, after a while. He was of an impulsive nature and generous sometimes. Besides, I could not help feeling that I had a moral claim upon him for a large amount." As usual, Zieber's fond hopes were dashed.

About eighteen months after returning to New York, I one day sought a private interview with Singer and Clark, at which I urged the injustice which had previously been done me and asked if they could not then, out of the great fortune they were making [this must have been 1854 when as we have seen no great fortunes were in fact being made], give me something, upon which Singer assumed to be very indignant, saying to me, "you're no inventor," which was a very silly remark to make upon the occasion, as I had never made any such pretension—and he turned his back upon me.

Nevertheless, being himself so fortunate as to be an inventor, Singer felt he could afford a certain social expansiveness toward Zieber, just so long as no unpalatable moral claims were pressed by his one-time backer. He would quite often invite Zieber to his home, "where I was treated with great kindness by Mrs. Singer and the family—but the sight of his . . . rich furniture, whilst I was now very poor, did not afford me much pleasure under the circumstances," the unfortunate fellow recalled. "The airs, however, assumed by people who had suddenly emerged from great poverty were sometimes curious and amusing. On these occasions he was very fond of talking about the business—and he once said, in presence of Mrs. Singer and all the children, 'If to anyone I owe my fortune, I owe it to Mr. Zieber.'" This, however, was something he could not be induced to admit anywhere outside the family circle. "He afterwards said, 'By God, I owe my fortune to myself alone,' which was true, in one sense, as he owed it, principally, to his great rascality."

There can be no doubt that Singer had acted in a rascally way, and not only toward Zieber. But while only the cynical or aggrieved would assert that rascality is an essential ingredient of business success, it would nevertheless be a strange world in which someone as weak and gullible as Zieber made a success of his affairs while a temperament as forceful and inventive as Singer's did not. Singer might fail or succeed, but he would do neither by half measures. Zieber was forever trying to justify his half measures.

It might be asked how I could remain in a situation so repugnant to me as the one I occupied must have been, after all that had pre-

viously occurred between Singer, Clark and myself [he now wrote],
to which I can only reply that on several occasions Singer inti-
mated that he would do something to help me, and it has sometimes occurred that men in early life committing robberies and
other disconscience [*sic*] which has forced them to make restitu-
tion. I hoped that this might be the case with Singer and Company, particularly after I had come back to return them good for
evil.

Alas, it *should* have been the case, but few realistic people could
seriously have believed that it *would* be so. In a way Zieber was lucky,
for in many instances the person who has perpetrated the wrong may
conceive an actual hatred for the injured party who comes back to
haunt him. This Singer did not do—indeed, he seems to have acted
with considerable cordiality toward Zieber, being as thick-skinned in
this as in other aspects of personal relations. Perhaps he lived so
completely in the present that he could never imagine himself in any
role other than the current one, or see how he could have acted any
differently. In Zieber's case, it was a question of dog eat dog in
business affairs; which is fine for the top dog.

In 1856 the Singers moved farther up Fourth Avenue to No. 395;
and in 1859, when Isaac was forty-eight and Mary Ann forty-one,
they bought No. 14 Fifth Avenue, just off Washington Square in the
most fashionable part of New York. Like Edith Wharton's Mrs.
Struthers, the Shoe Polish Queen, he "bought [his] house on Fifth
Avenue and issued [his] first challenge to society."

The Singers became more and more respectable. Mary Ann had
cards engraved "Mrs. I. M. Singer, 14 Fifth Avenue" and no doubt
spent her afternoons depositing them at the houses of acquaintances,
to return and find their cards duly left at No. 14. Tutors and piano
teachers were engaged for the younger members of the family. A
physician, Dr. William Maxwell, was paid to devote his entire attention to the family's needs. The household, as may be imagined, was a
lavish one. They kept six carriages and ten horses. Mary Ann had a
coachman at her sole disposal so that she or Isaac might each drive out
without inconveniencing the other. Ten dollars a day (worth about
$100 today) was allowed for groceries. Mary Ann could run up bills

from twenty-five to two hundred dollars in the comfortable knowledge that Singer would pay them uncomplainingly.

And yet, despite all this, society did not accept the Singers. Mary Ann, remembering those days, gave the names of the people who visited 14 Fifth Avenue: Mr. and Mrs. Dean, the parents of Julia Dean, Mr. and Mrs. Bosca, Mr. and Mrs. Meeker, Mrs. Walcott and her daughter, Mrs. Mead, Mrs. Noy, Mrs. Benari and others, Professor Mapes and Judge John W. Edmonds. Family ties were maintained: Charles Sponsler came to stay for several weeks at a time, and at one period two of Singer's sisters were living at the Fifth Avenue house. No doubt these were all worthy and delightful people, but they were not names with which society rang.

Certainly there was no social contact with Singer's partner in the business, Edward Clark. According to Zieber, Mrs. Clark, who, as the daughter of the state attorney general, had clear-cut views on her position in society, told a lady who visited her house that "she wished Mr. Clark would sell out, and leave the low occupation he was engaged in, and the nasty brute he was associated with." Zieber went on to say that "the extreme gentility affected by Clark's family troubled and annoyed Singer. They could not notice, in any social way, either him or his. They thought it soiled their fingers to come into contact with the money made by his patent sewing machine. Remarking to me, once, upon the thousands they were now able to luxuriate upon—the bills they ran up at Stewart's—he said 'Curse them! I am making them all rich.' " In this Singer was not entirely just, but there was certainly, by now, less love than ever lost between the two men. Perhaps Singer had got wind of Mrs. Clark's remarks. He was certainly indiscreet about his own feelings toward his partner. He once remarked to Zieber, after Clark had been involved in a carriage accident, that "if any thing serious should happen to Clark, by God, I will give the family a tussle for the property." ("I could not help wondering," wistfully adds Zieber, "whether in case he should have the opportunity of giving the family a successful tussle for the property, it was his intention to restore to me a portion of that of which I had been robbed.") On another occasion, Singer suddenly asked Zieber if he had ever seen Clark with his wig off. "No, I answered, rather amused at the oddity of his manner. 'Why?' I inquired. He replied, 'Because he is the most contemptible looking object I ever saw with his wig off.'"

This kind of cordial mutual hatred was the best reason in the world for not pursuing social relationships outside business hours. But Mrs. Clark's remarks indicate that there was in fact more to it than that. Could it be that in the country where a newspaper had once published a spirited defense of equality in reply to a Frenchman furious because he had had a suit of clothes ruined by an ill-directed gob of spit, a notorious American habit noted and deplored by all European visitors—

> John Tomkins, standing at his door,
> Spit on Jem Dykes, who passed before.
> "Sir," exclaimed Jem in furious fit,
> "When I pass by how dare you spit?"
> "Sir," replied John, with equal brass,
> "When I would spit, how dare you pass?"

—could it be that in the United States, land of equal opportunity and the self-made man, it was *not respectable to be in trade*? Was this the land of which Calvin Colton wrote, "Ours is a country where men start from an humble origin . . . and where they can attain to the most elevated positions, or acquire a large amount of wealth, according to the pursuits they elect for themselves. No exclusive privileges of birth, no entailment of estates, no civil or political disqualifications, stand in the path; but one has as good a chance as another, according to his talents, prudence and personal exertions"?

It was indeed; but it was that country grown some twenty or thirty years older, and in those few years New York society had begun to define itself. In the twenty years between 1840 and 1860, two roughly complementary changes were taking place. The first was bemoaned by George Templeton Strong the diarist, who was himself a member of one of the city's best families: "How New York has fallen off during the last forty years! Its intellect and culture have been diluted and swamped by a great flood-tide of material wealth . . . men whose bank accounts are all they rely on for social position and influence. As for their ladies, not a few who were driven in the most sumptuous turnouts, with liveried servants, looked as if they might have been cooks or chambermaids a few years ago"—or indeed oyster-packers' daughters turned traveling actresses.

Strong recognized that the new standards would be hard to resist as money assumed more and more importance in society. "Is it the doom of all men in the nineteenth century to be weighed down with the incumbrance of a desire to make money and more money all their days?" he asked himself, and concluded, "I supposed if my career is prosperous, it will be spent in the thoughtful, diligent accumulation of dollars, till I suddenly wake up to the sense that the career is ended and the dollars dross."

Nevertheless, society was determined to resist as hard as it could this new influx with its new values, and in order to do this the "good" families banded more and more tightly together—a reaction noted by Edith Wharton, possibly the sharpest observer of this society: "There was in all the tribe the same instinctive recoil from new religions as from unaccounted-for people. Institutional to the core, they represented the conservative element that holds new societies together as seaplants hold the seashore." The key word, of course, is "new." It is only when such distinctions have been drawn for a long time that they can be relaxed with any ease. Washington Irving had been able to mock the concept of a fashionable society in New York: "These fashionable parties were generally confined to the higher classes, or noblesse, that is to say, such as kept their own cows, and drove their own wagons. The company commonly assembled at three o'clock, and went away about six, unless it was in winter time, when the fashionable hours were a little earlier, that the ladies might get home before dark." This was a picture to be forgotten as quickly as possible.

This separation of the "good" families from the rest was to be given a certain tangibility by Ward McAllister, originator of such concepts as the "Patriarchs" and the "Four Hundred"—i.e. the number of those who "mattered"; this was the number of people who could be accommodated in the ballroom of society's then leading hostess, Mrs. William B. Astor. Mrs. Astor "mattered"; the Astors certainly had money before they had breeding—old John Jacob ate his peas and ice cream with a knife and was once caught wiping his hands on his hostess's sleeve—but they had been around for a long time and such indiscretions could be thankfully forgotten. The Vanderbilts, on the other hand, had taken longer to become established. In 1839 "Commodore" Cornelius Vanderbilt was already rich enough from his steamboat lines to go into a banking partnership with Daniel Drew of Drew, Robinson and Company, one of the most important banking houses in New York. But although he was at this stage worth more than half a million dollars—chicken feed compared with what was to come, but still a huge fortune in those days—the family was socially unknown and ostracized by those very Beekmans, Verplancks and Astors who were nevertheless so eager to do business with the Commodore.

By the time McAllister made his classifications the Vanderbilts were "in"—just: they were on the list of "swells," the ones who had to "entertain and be smart" if they were to be recognized. Old patroon families such as the Beekmans and Van Rensselaers were "nobs"; their position was such that it did not matter how they entertained, or indeed whether they bothered to entertain at all. This is not to say that even they were expected to be actually poor: nobody could comfortably be poor and remain in society, as any reader of such cautionary tales as Edith Wharton's *House of Mirth* will know.

McAllister did not really get into his swing until the 1870s, but by the late 1850s New York society had reached a pitch of exclusivity scarcely equaled in Europe. By 1860 magazines such as *Harper's* were keeping eager eyes on the doings of society and its adherents, and retailing the delicate distinctions drawn by the people who "mattered" for the delectation of those who didn't, as in this elaborate flowery allegory on the start of the season:

> The delicate descendant of the old shop-keeping or useful class, who had begotten a posterity more ornamental than themselves, began to stir with blossoming preparation. The flowering almonds and currants of several families, however refined and elegant they may have become, put themselves forth . . . The good old domestic names and places of those who have so long held a high place at court are still honored . . . But while thus all the fine folks are

dressed, the splendid newcomers of which I spoke are attired in the most gorgeous costumes. The wistaria is hung with purple raiment. It is even whispered in the choicer circles of the court— though the story sounds like the most dreadful court scandal, occasioned perhaps by jealousy of the foreign extraction of the illustrious strangers—that the wistaria dresses itself for the advent of summer with literally nothing but purple raiment . . .

It was enough to make the austere bones of George Washington rattle in his coffin.

America's Puritan tradition, however, had not failed to leave its mark even on this section of society. The last allusion in the *Harper's* extract was probably to August Belmont, the mysterious, foreign, probably Jewish banker who made such a determined assault on the bastions of New York society at this time. Belmont was never accepted quite without question despite—or, more likely, because of —his enormous displays of ostentatious good living. For among the external signs of genuine good breeding—dinner party guests were advised against "shaking with your feet the chair of a neighbor"; "the word 'stomach' should never be mentioned at table"; if a lady should make an "unseemly digestive sound at dinner" or "raise an unmanageable portion to her mouth" one should "cease all conversation with her and look steadfastly into the opposite portion of the room," and so forth—one of the most important was that one should not be ostentatious.

In his study of the fashionable New York Jewish world of that time, consisting mostly of bankers and financiers and running roughly parallel to, but rarely intermingling with, the Four Hundred, Stephen Birmingham points out that "one worried about being 'showy,' and spared no expense to be inconspicuous." In 1859, a no-doubt-surprised English readership, unused to such austerity, was informed that "the style of everyday living among even the wealthiest people is very simple and inexpensive. But little wine is drunk in the domestic circle; and plain English cookery is alone usual. Eating and drinking, *en famille*, is a mere operation of appetite, without any social feeling connected with it; and the more quickly and least expensively it can be performed the better."

For the Vanderbilts, Rockefellers and Carnegies who were to necessitate the coining of a new phrase—*multimillionaire*—this insistence that public show by the wealthy was vulgar and should be eschewed was not a great deprivation. In Europe one made money in order to live the good life, possibly crash society, and generally fulfill one's fantasies. European culture had long since developed a large number of ways in which money, once acquired, could agreeably be disposed of; such milieux still formed the background of almost all literary works. The same was not true of America. It was not that European capitalists were not as concerned as American ones with the economy they were creating; simply that this was never their sole preoccupation. Indeed, business has never become in Europe the overriding preoccupation which it still is in America. At the time of which we are speaking, moreover, Europeans were expanding their economies while Americans were virtually creating theirs. The obsession with business and money-making for its own sake which now appeared and which was to become so characteristically American seemed both logical and patriotic. On the one hand all the pressures of the new country (as distinct from New York society) put an enormous premium on entrepreneurial skills: business was a worthy occupation in itself. On the other, opportunities for conspicuous consumption were in many cases simply unavailable (though this was a situation which would change radically during the next half century). The luxuries and sophistications which formed the core of the way of life of the European rich did not yet exist in America—arguably, many of them have still not arrived.

This lack of opportunity for recreation was in many parts of America still bolstered by law. In almost all states, games, recreation of all sorts and public entertainments were banned on the Sabbath and frowned on during the week, while some states actually had laws enforcing church attendance on Sunday. The primitive need for ritual atonement is, of course, catered to in a variety of religions, and for the successful businessman, the application of the Puritan ethic in private life may well have served this purpose. Many a dubious business conscience and reputation was salvaged by regular appearances in church on Sunday. He could then turn with increased zeal to a process of moneymaking whose only visible goal was to make more money. It was a new and totally American phenomenon. What such men were hatching was in effect the philosophy of the conglomerate, where the object is to increase aggregate wealth in order to acquire more wealth rather than in order to spend it or to achieve excellence in any one field of endeavor. The early millionaires were mostly, so to speak, one-man conglomerates. They would put their cash where more was to be made rather than because of any abiding interest in railroads or, let us say, sewing machines. Indeed, many of the wealthiest railroad barons made their fortunes purely from rigging the markets, without organizing the laying of a yard of track or the construction of a single sleeper.

Singer, despite his declared interest in "the dimes, not the invention" (which nevertheless failed to conceal a passionate interest in his machine), could in no way share these attitudes. His reasoning was simple. He had been poor and had not liked it, and now that he was rich, if there was one thing he was going to do, it was enjoy himself. He was prepared to work hard and pull his weight in the business: in 1857 alone, he assigned to Clark a half share in twelve patents, under the terms of the partnership. But unlike most of his compatriots he also had a well-developed penchant for pleasures of all sorts—as evidenced by his continuing enthusiasm for that frivolous nonessential, the theater. (It is worth noting that almost all the leading actors and actresses in the States at this time were European imports.) Furthermore, for Singer the mere fact of enjoying something had never been enough. He was a natural showman, he liked to be seen and heard, and his pleasure was doubled if he could be seen and heard to be enjoying himself. Melodrama was the key at which his life was habitually pitched, and the melodrama of riches is extravaganza. I. M. Singer was already known for his sewing machines. Now he would become even better known for his style. Nothing could have been better calculated to draw down the disapproval of the world in which he lived.★

★ This is not to say that none of the other nineteenth-century millionaires ever broke out and enjoyed themselves, but by the time they did so they were both much richer and better established in society than Singer was in 1859. For instance, when "Commodore" Vanderbilt became enamored of the

He began by designing himself a coach. The Singer family very much liked driving in the park and did so nearly every day. Unfortunately, although there were always a number of carriages at their disposal, none was big enough to contain the whole brood with their friends and the accompanying servants. Isaac now set about constructing such a vehicle. It was patented (no. 25,920) on 25 October 1859—one Singer patent in which we may be sure Clark did not have a fifty percent share. The inventor stated gravely in the specifications:

> The object of my invention is the production of a carriage, mainly intended for family use, which shall have, within a smaller compass than any other known plan and within the main body, a larger number of seats so arranged that the persons sitting on them can conveniently converse, and to obtain this result in such a manner as to admit of placing the center of gravity of the whole structure sufficiently low for convenience and safety in traveling on common roads, and to admit of symmetry in the general structure, while at the same time inside seats for children and servants and suitable receptacles for all the necessary conveniences of travel are provided, as also an increased number of outside seats.

It is an endearing concept. He was proud of his family—indeed, he was obviously very fond of his children—and wished to be seen in public as a family man. The needs of children were specially taken into account: a special section of the coach was designed for them.

> At the back of the main body and attached thereto is a depressed coupé body . . . The coupé is provided with a middle door in the back, and with seats on each side, the space under one of the said seats being suitably arranged as a water closet, and the other as a receptacle for baggage &c. From the coupé openings are made . . . through the back of the main body to the spaces under the two back elevated seats, one for storing baggage, and the other for an extension of one of the side seats of the coupé to form a child's bed,

spiritualist and free-loving feminist Tennessee Claflin, and asked her to marry him (in 1868) he was immeasurably rich and his children and grandchildren immovable pillars of society. He was then seventy-six, and it was the first time in his life he had ever had the time to think of any pleasure but business.

or if desired both may be arranged for beds, or both for the reception of baggage, &c . . . By the arrangement and general construction of the main body and seats therein, I am enabled to obtain by a very slight increase of size ample room for nine, and if necessary ten persons, all so situated that they can converse freely and none of whom, if limited to nine, will be required to ride backward; while in the coupé at the back and in communication therewith there will be ample room for children and servants and all the conveniences of travel.

This amazing vehicle was indeed built, and the family drove out in it very frequently. The *Scientific American* described it as being "after the style of a Russian nobleman's equipage, and resembling more than anything else a continental diligence of olden times." The *New York Herald* was suitably astounded by it:

The most remarkable equipage that travels Bloomingdale is the turnout of I. M. Singer—a regular steamboat on wheels, drawn by six and sometimes nine horses, three abreast. The wheel horses are two large greys with a bay between them, led by a light-coloured cream between two sorrels. When Mr. Singer turns out with nine horses, three cream-coloured ones are placed as leaders. The carriage is a monster, having all the conveniences of a modern brownstone front, with the exception of a cooking department. It weighs about 3,800 pounds and will seat, inside and out, thirty-one persons. The inside is arranged with convenient seats for full-sized persons with a nursery at the back end for the nurse and children, with beds to put the dear ones to sleep, and all the other necessary arrangements. This apartment is also used as a smoking-room when there are any gentlemen in the party who wish to enjoy their Havanas. There are seats on the outside for the accommodation of sixteen persons, these furnishing means for carrying a small band of music, with guards enough to keep off all outside barbarians. Under the front of the carriage is a baggage compartment large enough to deposit the luggage of a dozen or more persons. The body is painted a canary bird color, edged with black. The footman is always dressed in great style, whilst the driver appears in ordinary attire.—Mr. Singer has three other carriages, and sometimes appears with three horses abreast before an ordinary coach. He has invested in carriages alone six thousand dollars, of which three thousand is for his large sociable, or traveling carriage. This gentleman has ten horses in all, which have cost him ten thousand dollars. Whether his eccentric turnout is intended for speed, comfort or advertisement, the reader must judge.

When not riding in this omnibus, Singer amused himself (so the weekly *Family Herald* reported) by driving through Broadway at great speed sporting a dashing five-in-hand "unicorn team" of three horses abreast and two leaders, at great danger to pedestrians.

Such goings-on appalled Clark (summed up by his obituary in *Harper's* in 1883 as "a very charitable man, but quite unostentatious"). They were the height of bad taste, the very acme of the sort of thing one absolutely did not do if one wanted to become a member of polite society. No wonder the Clarks were eager to dissociate themselves from such vulgarity! What was more, Singer was setting a dangerous precedent. If one partner could display his wealth in such a showy manner, people might well assume that both the firm and the other partner were equally wealthy. This was not an impression Clark wished to encourage. In April 1861, when the country was building up to the start of the Civil War, he was to write to his partner, then in England:

Business is pretty much at a standstill. All are full of war. Our men are going off as soldiers pretty rapidly and to them their arrears have been paid in cash. The others have their pay half in machines or wait. In pride and feeling, I am suffering for all the large public show of wealth you made in 1859 and 60. It was industriously spread abroad that the firm was rich. Now all who are rich are expected to be patriotic and to give liberally . . . I am called on many times a day to subscribe and am obliged to refuse.

By that time, however, Clark had something a little more substantial than bad taste with which to reproach his partner—something which, while no doubt paining him deeply, cannot but have afforded him a certain satisfaction. There is always some pleasure in having one's worst suspicions amply confirmed.

* * *

The year 1860, his forty-ninth, was to prove an eventful one for Singer, not so much in business as in personal matters. For the past nine years, since I. M. Singer & Co. had been established in New York, his life had been jogging along on an even keel to all appearances, his family growing as his circumstances steadily improved. If the Singers were not invited to dinner by the Van Rensselaers nor the Vanderbilts, they nevertheless had their own friends and a satisfactory way of life established—at least as far as Isaac and Mary Ann were concerned. It might not have seemed quite so unexceptionable to their children, especially their daughters—Voulettie was twenty in 1860, an age when the social standing of one's parents can matter very much—but they could always rely on the phenomenon of society's notoriously short memory as to how a fortune was actually acquired, especially once the original undesirable is safely dead. Edith Wharton, cousin of Newbolds and Rhinelanders, was able to mix with the grandson of the arch-swindler and social outcast Jay Gould. But at the time, it might well have seemed that the events of 1860 had put paid once and for all to any such hopes.

The first of the year's events was that, in January, Isaac and Catharine were finally divorced. It might be supposed that Catharine, having such total and public grounds for action, finally lost patience and divorced Isaac; but this was not the case. It was he who divorced her on the grounds of her adultery with Stephen Kent. On 23 January Catharine's lawyer, a Mr. Lambier, called on her and told her that he could obtain $10,000 for her from Singer if she would consent to have a judgment of divorce entered against her. Mr. Lambier advised her to take the money while it was offered, as apparently Mr. Singer was in "falling circumstances" and the offer might not remain open for long. Acting upon this advice she agreed to receive the money. In return she swore before a referee that she had committed adultery with Kent and named times and places. Kent confirmed this testimony. In these circumstances the judge could not but grant a divorce *a vinculo*; adultery was still the only grounds upon which divorce was possible in New York State.

It was patently only the threat of funds being imminently cut off altogether which led Catharine to agree to this otherwise quite inexplicable move. She was still receiving money on a regular basis from her husband, and as his circumstances improved so, naturally, did hers. It was only if he were to crash, as was now hinted, that she would be better off with a divorce and cash in hand.

From Isaac's point of view the matter is equally hard to explain, especially if one takes into account his subsequent behavior. He had been married to Catharine for thirty years and this had if anything enhanced his freedom of action with other women, since the one thing they could never expect him to do was marry them; they took it or left it on that basis. Mary Ann, the chief though not the only sufferer from this state of affairs, must long ago have become resigned to her lot; at any rate, she seemed to have accepted it. Surely the very last thing Isaac wanted to do was bring the whole question to the surface again, as must inevitably happen if he got a divorce? Irregularity in no way bothered him, and he was getting along very nicely, thank you, as things now stood. Perhaps the key to the affair lies in the testimony of Lillian Singer, daughter of Isaac and Catharine, who said she was present on an occasion when Edward Clark offered to pay her mother $10,000 if she would consent to have a decree of divorce entered against her. Clark, of course, was the one person to whom the chronic irregularity of Singer's affairs now mattered—and they mattered very much. Quite apart from his concern with respectability, at which the immorality of Singer's arrangements must always have nagged—for Clark must have known about Catharine and, therefore, known also that Mary Ann was not the real Mrs. Singer—he was a very pious man and must have been genuinely distressed by the apparently ir-remediable bastardy of Mary Ann's brood. As the business grew, and with it his, Clark's, own wealth and prestige, he came to feel increas-ingly that Singer's bad ways were reflecting on both the firm and his own social standing. It seems probable, then, that it was Clark who persuaded Singer to take this step, assuming that he would thereupon marry Mary Ann. It was a Clark-like touch to arrange affairs so that the legal innocence and thus the official good name of his partner should be maintained by the divorce agreement. Such details of course redounded on the respectability of the firm. Unfortunately, Clark had forgotten that you can lead your horse to water, but you can't make him drink.

The news of the divorce came through when Singer and Mary Ann were in Baltimore attending the hearings of an important legal suit in which a party named Walmsley was attempting to prove that Singer was not entitled to any of his original patents. Various witnesses were brought in to testify that they had invented Singer's machine, or its equivalent, before he had; Orson Phelps appeared to say that it was he, and not Singer, who had done the design work on the original Jenny Lind machine. (This was the last occasion on which Phelps and Singer were ever to meet). Clark had been much occupied with the finding and preparation of witnesses for this trial. "Dear Sir," he wrote to Singer in Baltimore on 16 January—a form of address which encapsulates the coldness of their personal relations, considering that they had, by then, been partners for nine years— "I have had a long conference with Mr. Lafetra today and think I shall be able to get him to attend as a witness. He is inclined to know a little too much but if he is confined to swear to what he did when examined by Mr. Potter, his testimony should help us decidedly. Yours truly . . ." This was typical of the kind of painstaking ground-work which had to be done before any important patents trial: Lafetra was supposed to have invented a machine before Singer, and Clark had been maneuvering to pull his testimony into a form favorable to the defense rather than the prosecution. It was the kind of thing at which he excelled and with which Singer could never be bothered.

Singer and Mary Ann stayed for the duration of the trial, about six or eight weeks, at Barnum's Hotel, where they had a parlor and a suite of rooms. They were, naturally, known as Mr. and Mrs. Singer, and as such they were visited by, and paid visits to, among others, Mr. Latrobe, Singer's lawyer in the case, as well as Mary Ann's family. But the arrival of the divorce papers brought Mary Ann's attention more forcibly than ever to the fact that they were, after all, *not* Mr. and Mrs. Singer but, in the eyes of the law, Mr. Singer and Miss Sponsler. When the papers arrived in Baltimore, Singer actually handed them over to Mary Ann for safe-keeping. He could hardly have failed to anticipate her next question: now that Catharine, at long last, was out of the way, when would he fulfill his frequently-repeated promise and marry her? She can scarcely have anticipated his reply, which was that he did not intend to marry her, since if he were to do so, "she would have him in her power." There was no more to be said.

Against such a refusal, Mary Ann had no weapons. She had no means of forcing him and the more fuss she made the more likely it was that the whole story would be made public and her position would be seriously undermined. If, on the other hand, she kept quiet, life could go on as before. No doubt Singer had banked on this logic. Mary Ann's frame of mind can be imagined. She must have felt at the same time furious, nonplussed and deeply hurt. On the strength of a promise she had literally given him the best years of her life, and borne him ten children. Now he told her that he would not keep his promise because, not to put too fine a point on it, he did not feel inclined to do so. She left Baltimore, saying she was worried about the children but more probably because she simply wanted to get away from him, and returned to New York. When the suit was successfully finished, Singer went on to Philadelphia, whence he telegraphed Mary Ann to join him. She did so, and later they returned together to 14 Fifth Avenue, where life resumed more or less its old course, although we may imagine there were undercurrents of tension and discontent which went studiously ignored by the master of the house.

Such a *modus vivendi* could perhaps have lasted indefinitely; but it was given no chance to do so. The end came as a result of one of those sudden and sensational incidents which seem like mere bad luck at the time, but which hindsight endows with a certain inevitability. As Mary Ann put it, "We lived together until August 7, 1860. A difficulty occurred then."

The difficulty was a spectacular one. Mary Ann, who was driving along Fifth Avenue in her own carriage, met Singer driving in the opposite direction in an open vehicle with one Mary McGonigal seated beside him. Mary McGonigal was not entirely unknown to Mary Ann: she later said that she had "seen his companion in the Singer Company's Philadelphia offices more than once, but not with Mr. Singer. She had heard rumors, however, of the relations between them"—rumors which were now to all appearances confirmed. Driven thus by a double fury, Mary Ann began to scream. She screamed at the top of her voice—screaming nothing in particular, as she later said: "I used no language upon this occasion"—just screamed until she caught the attention of the occupants of the open carriage, not to mention half the current population of Fifth Avenue. She then turned her carriage down a side street and into Third Avenue and drove home.

By the time she arrived at her house she must have realized that her conduct, although doubtless it had considerably relieved her feelings, had been most unwise. By making such a public scene she had ensured one thing: her present way of life could not go on; things could no longer be decently covered up and something or someone would have to give. But what position was she in to dictate terms? The sight of Singer awaiting her in a fury at No. 14 can have done nothing to reassure her. He, as might be expected, had been thoroughly enraged by the scene and he now administered what she later termed "a brutal and bloody assault." Mary Ann was probably half expecting this as he had beaten her several times in the past— the previous year, for example, when he had been angered by the sight of a box of matches left uncovered in the bedroom. On that occasion, Voulettie had intervened to calm him down, whereupon he set about her as well, until she and her mother both lost consciousness. He had become so worried about their condition on that occasion that he called Dr. Maxwell, and the two women were confined to their beds for some days. At that point, Mary Ann had actually left Fifth Avenue and gone to stay with a friend, Mrs. Megary, who kept a boarding house at 83 Clinton Place, but she had come back a few days later.

This time, however, she decided she had had enough. In its News of the Week, the next issue of *Frank Leslie's Illustrated Weekly* carried this item: "Mr. I. M. Singer, the great sewing machine manufacturer, who resides at No. 14 Fifth Avenue, New York, was arrested on complaint of his wife, who accuses him of having beaten and choked her in a violent manner. He was put under bonds to keep the peace for six months." At this time, *Leslie's Weekly* had one of the largest circulations in America, selling 150,000 copies every week and read by 500,000. No disgrace could have been more public. On 19

September Singer fled, taking the boat* for Europe with Mary McGonigal's nineteen-year-old sister Kate under the names of "J. W. Simmons and Lady." On arrival in London Singer took up residence in Cornhill, near Cheapside, where the company had an office.

Clark was beside himself. So much for his dreams of respectability! The final straw came when a bank refused to advance the firm $3,000 on account of Singer's moral turpitude. "Now all this is exceedingly annoying to me as well as disastrous to our business," he wrote in a fury to his partner.

> I hardly dare speak to any old friends when I meet them in the streets. The firm of which I am the active manager has been publicly accused of keeping numerous agents in various cities to procure women for you to prostitute. And although this is an infamous falsehood, yet it is mixed up with so much truth that it would be disgraceful to bring into the light of a public trial, that neither I, who am most injured in money and reputation, nor the agents at the branch offices who are outrageously slandered, dare to appeal.

It is hard to know whether it was Clark or Mary Ann who was more shocked and startled at this sudden turn of events, with all its accompanying hints, revelations and publicity. But what exactly was going on? How far did the publicity correspond to the truth? And who, for a start, were the sensational McGonigal sisters?

Of one thing we may be sure, and that is that no rumors could possibly have exaggerated the truth. For the story of Singer's secret life was indeed an astonishing one.

At the time when Mary Ann made her scene on Fifth Avenue, thus letting all the cats out of their bags, Mary McGonigal had known Singer for nine years and was the mother of five children by him, the first, Ruth, born in 1852, and the youngest, Charles Alexander, born in 1859, named (did she know it? perhaps she did) for the little boy who died in 1854. Singer had taken a house for her and this family at No. 70 Christopher Street. Here she resided under the name of Mathews; Singer was known to their friends as Mr. Mathews, and all their five

---

* This was in fact the first west–east crossing of the Atlantic by I. K. Brunel's *Great Eastern*.

children carried that surname. Mary's younger sister Kate, the one with whom "Mr. Simmons" fled to Europe, also lived in the house.

This was not all. It emerged that Singer was supporting yet another family in Lower Manhattan. In 1851 he had had a daughter by yet another Mary, Mary Eastwood Walters. For this family he used the name Merritt; this Mary was known as Mrs. Merritt, and her daughter was called Alice Eastwood Merritt. For his Merritt family Isaac also took a house, No. 225 West Twenty-seventh Street. Thus by 1860, Isaac had fathered and recognized eighteen children, of whom sixteen were alive, and could claim to be supporting them all in a very reasonable style. That is not to say they all received the same treatment, for at this stage there is no doubt that the most favored of his families was the "official" one, Mary Ann's, which lived in such lavish surroundings in Fifth Avenue and drove out in their canary-yellow omnibus. However, he seems to have had a remarkable capacity for keeping all his brood in mind—this was borne out by the terms of his will when he died—and he kept a watchful eye on all their careers. William, who was now twenty-seven, was employed in the factory, as was Gus; both these sons had of course been associated with the venture from its earliest days. Voulettie was to marry the up-and-coming young man of the firm, William Proctor, but at this point she was still living at home with her mother, to whom she gave considerable support.

The situation, when one considers its implications, was a truly amazing one. What emerged in 1860 was that virtually ever since his arrival in New York, Isaac had been running three families; not counting the fourth, the legitimate one, with which he had least contact. The dates of birth indicate that he must have taken up with Mary Walters, "Mrs. Merritt," as early as 1850 (when one might have supposed he was wholly occupied with the invention and development of his machine) and with Mary McGonigal, "Mrs. Mathews," in 1851. No wonder Zieber grumbled that he was always drawing so much money out of the firm—"at least three times as much as I did." He must have needed every penny, especially in the early days.

What inkling, one wonders, did each of these three Marys, Sponsler, McGonigal and Walters, have of her own true status? Was

"Mrs. Merritt" ignorant of "Mrs. Mathews"? Could either of them have been unaware of the existence of "Mrs. Singer"? As for the characters of Mr. Merritt and Mr. Mathews, for whose benefit were they assumed? Did they ever obscure the reality of Mr. Singer for their respective "wives"? Were they purely a front for friends and acquaintances? Or were they a psychological device allowing the *père de famille* to assume more easily the role of the day? Were they perhaps a dramatic aid, enhancing for him the drama which had always been such an essential ingredient of his life and which, we can now see, he had by no means abandoned when he quit the stage to assume the role of inventor and businessman? It is easy to see how he must have enjoyed becoming Mr. Merritt and then Mr. Mathews as a change from playing the constant part of Mr. Singer the respectable sewing machine manufacturer. At least, since all these "wives" were named Mary, there was no danger of waking up with the wrong name on the tip of his tongue.

Although there is, naturally enough, little reliable information on these topics, there are a few indications as to who knew what about whom before the "difficulty" took place on Fifth Avenue, thus blowing the whole carefully constructed edifice sky-high. Mary Ann knew about Catharine; she had known about her from the first, of course, and she may well have met William when she went, as from time to time she did, to the factory. She had heard "rumors" about Singer and Mary McGonigal, and indeed it seems unlikely that these were the first such rumors she would have heard about him; it will be remembered that similar rumors had begun to accumulate while he was still living with Catharine. Being a woman of perception and intelligence (which according to all reports she was) she cannot have imagined that he stayed faithful to her during their repeated separations, from 1837–39 when she was with her mother, then when he was in Boston with Phelps, and while he was traveling around the country on business. His own behavior to her when they first met must have given some indication of the way he was likely to go on. What she presumably did not imagine was that there was another woman in a precisely similar situation to her own—that is, the mother of a large family of Singer children. Had she known this, she might have pressed him even harder to marry her, since the existence of these other families would severely have undermined any confidence she may have had in being unquestionably the *real* Mrs. Singer, by right if not by law. The one totally intolerable thought for Mary Ann must have been that Singer might end up actually marrying someone else, after all she had been through with him.

Of Mary Walters nothing is known, but it is hard to imagine that Mary McGonigal was ever deceived by the "Mr. Mathews" front. If she had indeed been at the Singer Company's Philadelphia offices "more than once," as Mary Ann asserted, she must have known that her lover was none other than Mr. Singer himself. She must certainly therefore have known about Mary Ann, and possibly about Catharine. Although the knowledge of her lover's identity must have induced a justifiable sense of financial security, this obviously did not weigh heavily enough with her to outface the scandal which now overtook her and her children. Determined to get herself and her family as far away from it all as she possibly could, she took them to San Francisco, where under the name of Mathews she eventually eked out a living by taking in sewing! Here she was traced by Singer after a long search when he finally returned from Europe. (It seems that Mary Ann never in fact connected Mary McGonigal with "Mary Mathews," since in a complaint she later made against Singer she named these ladies separately.)

All Singer's acquaintances agreed that he was singularly energetic, and if proof were needed, the fact that he had for nine years managed such a life as well as remaining active in the business would go a long way toward providing it. How did he possibly find time for it all? Were nights or days per week allotted in accordance with the number of children involved? One journalist put forward the theory that Singer spent alternate nights with each of his four families, but Mary Ann contended that "Singer was never out at night, but stayed at home with me, except when he would go to the theater with his family." Of course he could always have used the time-honored subterfuge and told her he was going out of town on business—something less easy to check in the days before telephones—but there was presumably a limit to the number of times any excuse could be used

over the course of nine years. There remained the possibility that it
was during the day that he became Mr. Merritt and Mr. Mathews,
and this possibility is given credence by his coachman, Edward de
Longue, who recalled that he

> frequently drove the carriage for Singer and . . . had known of his
> driving out with other females; sometimes he would drive to No.
> 225 West Twenty-seventh Street and there pick up a young woman
> and at other times he would take up one at No. 70 Christopher
> Street; he often used to take up females who would be in waiting
> for him on the corners of streets, with school books in their hands;
> he would nearly always give me some change for myself on these
> occasions; I was at these times ordered to drive to the Central
> Park and back again.

A quick one indeed and not something to foster the pretense of
normal family life! The item about the "school books" strikes a particularly odd note: was it intended to indicate that Singer had a
perverse preference for little girls? Certainly his choice of more permanent mistresses and wives would not seem to show any such leanings,
apart from Catharine—who at fifteen was nevertheless only four
years younger than her nineteen-year-old groom. Mary Ann was
eighteen when they met; Mary Walters was twenty-seven; Mary
McGonigal was twenty-four; and Isabella, his last wife (whom at this
point he had not yet met) had already been married once before.
Perhaps the girls with school books were merely some of his many
daughters?

★　★　★

During the ten years 1850–60, Isaac's life had been complicated but it
had also, amid its various convolutions, been under control. To the
passing fly the web might have seemed hopelessly entangled, but
the spider in the middle had everything worked out so that, despite
the occasional tremor, it remained intact. Now, however, the delicate
fabric was broken, and for the next three years the dramas of Isaac's
personal life were acted out in various courtrooms in and around New
York. They attracted less attention than might have been the case
in less troubled times: the Civil War was raging and this naturally

occupied the headlines. Meanwhile the main protagonist flitted back and forth between America and Europe, apparently quite unabashed, while Edward Clark battled to maintain the business on an even keel and preserve it, so far as he could, from the taint of scandal.

The Singer Company letterbooks give some picture of Isaac's progress on the other side of the Atlantic. He was obviously living in some style, since the London agent frequently reported to Clark at this time that his remittance to New York would be less than anticipated because "I gave, or rather sent Mr. Singer £100 . . . and he has ordered another £100, as well as £100 on or before the 20th, this must be supplied to him." Such news must have been particularly galling for Clark, since it was just at this point that large subscriptions were being demanded of him to help finance the Union armies; moreover, as the war progressed, cash was (as we have seen) increasingly tight and the firm came to rely heavily on its remittances from overseas.

Meanwhile on Fifth Avenue (where she was still living) Mary Ann decided that her affairs and those of her children had better be set in some kind of legal order. For this she could scarcely be blamed; in less than a year her world had shattered around her ears and she was in the most insecure position possible. On past form it seemed likely that Singer would not abandon her financially, even though he had done so physically; and if, as Mrs. Singer, she had drawn on the company during the preceding ten years for her day-to-day expenses, there was perhaps no reason why she should not continue to do so. Nevertheless, it was now public knowledge not only that she was not Singer's legal wife, but that he kept other mistresses. Although the full details of his personal affairs were not generally known until after his death, we may safely assume that they were by now more or less familiar to Edward Clark, and possibly to other confidants, such as George Ross McKenzie, within the firm, so that Mary Ann may well have felt diffident about presenting herself at the office. They at least would be aware that she no longer had any especially privileged claim on Singer, since at least two other women stood in precisely the same relation to him as she did. Her concern, therefore, was to prove the legal as well as the moral priority of her claims, and she now set about doing this. Her method was ingenious: she decided to sue him for divorce on grounds of adultery, claiming that although she had never formally been married to him, the fact that he had lived with her for seven months after his divorce from Catharine established her as his common law wife.

Mary Ann set out her accusations in a detailed complaint which was finally presented to the court by her lawyer, Mr. Fuller of the firm Abbott and Fuller, in December 1861. This document gives the most hair-raising picture of her domestic life with Singer. In it she alleged

> That, during her whole married life, she has received from her husband the most cruel and inhuman treatment, and that his conduct toward her has been such as to render it unsafe and improper for her any longer to cohabit with him; that he has repeatedly beaten and choked her to insensibility, frequently forcing the blood to flow in streams from her nose, mouth, face, head and neck, and the recital, to her family and friends, of his unparalleled atrocity and savage barbarity toward herself and children finally forced her to make a criminal complaint against him and cause his arrest for a brutal and bloody assault.

She also set out his adultery in detail, naming not only Mary and Kate McGonigal and Mary Walters but also a

> Mrs. Judson, with whom, at various times from 1857 to 1860, and at divers places in the cities of New York, Chicago and elsewhere, he frequently committed adultery; that the said Mrs. Judson was one of the operatives in the establishment of the defendant in the city of New York, and for the past two years has been an attendant in the Chicago business office of I. M. Singer & Co. . . . That the plaintiff is also informed that he kept as his mistresses . . . Ellen Brazee and Ellen Livingston, with each of whom, at various times and in divers places in the city of New York and elsewhere he has frequently committed adultery, and by whom he has had illegitimate children. That . . . he now keeps as his mistress one Lucy ___, a young English girl, whom he seduced in England and who followed him hither, shortly after his return to this city, in July 1861, with whom he now lives in open adultery at No. 110 West Thirty-seventh Street, New York, where he has a furnished

house . . . That the defendant is a most notorious profligate and
——, and that a more dissolute man never lived in a civilized
country; that he is in the constant habit of seducing all his female
operators who will submit themselves to his base desires, and
whom he employs in his establishments in this country and in
Europe with special reference to making them the victims of his
brutal lust . . . [lastly] That the defendant has threatened the plaintiff and her lawyer, to his face, with the "God damnedest beating
anyone ever had in this world" if any legal proceedings for a divorce should be commenced against him by her.

Few more lurid documents can have come before the court that year
(or indeed any other year).

    Such documents always present problems for persons reading
them who were not acquainted with either of the parties involved and
did not witness the home life and scenes they describe. It seems very
probable that all the allegations Mary Ann made were strictly true.
Certainly Singer never made any effort to deny the ones regarding his
adultery; and the rumors about his seducing his operatives (politely
described by Edward Clark as "infamous") seem on the contrary to
have been mere sober recitals of fact. As for the violence, that too
seems well in character. His brutality and brusqueness had been
observed by almost everyone who met him, and his first reaction
when faced with a situation he disliked was always to try to intimidate his opponents: an easier proposition when he was faced only by
Mary Ann than by (to give the nearest example to hand) Mr. Fuller,
her lawyer. This bullying was the obverse side of that energy and
drive which were also frequently noted; it was a judicious combination
of the two which had got him from the pinched poverty of Oswego to
a position where he could and—to the horror of society and above all
of Edward Clark—did satisfy his every desire.

    But of course that was not the whole story. Time and again when
considering Isaac Singer we are faced with this paradox: that he is
presented as a black and brutal villain in situations which could not
possibly have arisen if this had been the whole truth. In fact most of
the personal records we have of Singer are biased in that they were
written by persons he had wronged, who wanted to put their side of

the story in order to get their fair share of the loot, and were therefore intended to show him in the worst possible light. This was why
Zieber wrote his memoir and why Mary Ann framed her complaint.
In fact, Zieber's record of Singer is, despite this, not altogether black;
he shows a number of reasons—generosity, impulsiveness, the fact
that he could be an amusing and persuasive companion—why Singer
was able to persuade members of either sex to take up with him,
possibly against their better judgment. In addition to these personal
qualities, potential business partners would be aware of his energy,
inventiveness and flair for salesmanship, while women were no doubt
bowled over by his exceptional good looks (still well in evidence even
at the end of his life) and that confidence in his own "star quality"
which had characterized him from the first and which is such an important ingredient of this type of success. As for Mary Ann, there were
occasions—notably in the contest over his will when Singer was dead
—when, in trying to achieve a different end, she painted quite a
different picture of their married life: one in which family and friends
visited, they drove out with the children, grieved together when the
babies died, and led a quiet and comfortable life together over a
number of years. According to Charles Sponsler, who often visited
them, "Mr. Singer seemed to treat my sister kindly." The truth, as
usual, probably lies somewhere between the two versions.

Whether or not it contained the whole truth, however, Mary
Ann's 1861 complaint was clearly dynamite. Such revelations would
be sensational today, especially if they concerned a well-known public
figure; a hundred years ago these things were barely dreamed of by
those respectable people who so eagerly scanned the pages of the
Family Herald and the *Police Gazette*. Singer, now back from Europe
and living with his Lucy while Mary Ann still remained in Fifth
Avenue, put in an answer denying the marriage but admitting the
adultery, a paradoxical position. The judge in the case granted Mary
Ann her divorce and awarded her $8,000 a year alimony, the largest
such sum ever yet awarded; but his order was never executed and the
divorce was never made official. Singer became alarmed and suggested
a compromise, which Mary Ann for some reason accepted. Perhaps he
was threatening to contest the award. In any case, seven days after

the decision in her favor, "Isaac M. Singer of the first part" and "Mary Ann Singer of the second part" executed an instrument which said that "whereas, it is deemed desirable by both of the said parties to avoid the annoyance of further litigation of said action, and the probable scandal and disgrace to their children of probable trial of issue," it was agreed that he would procure and convey to her a pleasant house in some respectable part of New York, the title to which, on her death, should be vested in her children. She was also to be allowed fifty dollars a week alimony. This instrument did not debar her from pursuing her divorce suit, but should she elect to do so, it must be at her own expense and without any claim against Singer. A house at 189 West Twenty-eighth Street, between Eighth and Ninth Avenues, was duly found and furnished for Mary Ann, this task being performed by George Ross McKenzie, whom we last met lending the company five thousand dollars out of his own pocket and who now executed all Singer's confidential business.

Mary Ann's lawyers advised her that this was a generous settlement and they felt she would do well to leave it at that. Nevertheless, she was determined to obtain her official freedom—perhaps because this would be the only satisfactory way of proving that she had indeed been married. Two days after the arrangement about the house had been concluded she visited her lawyers once more, handed over a check for five hundred dollars and announced that she wished to continue with the divorce suit. After a time, although no decree appears to have been obtained, she was given to understand that she was free to marry whom she pleased, without prejudice. The next month, June 1862, she made a trip to Boston and there got married, under her maiden name, to one John E. Foster. He gave his age as twenty-five and she gave hers as thirty-one, although she was in fact forty-five at this time. That is perhaps only too easily understandable, but what is stranger, in view of all her recent activity, is that she also said this was her first marriage. This was of course strictly true in that it was the first time she had been through a marriage ceremony, but she had spent the past six months earnestly trying to show that it was *de facto* untrue. It was a slip she was to regret.

It is unclear how long Mary Ann had known Foster, who was a

passenger agent for the Erie Railroad. She appears to have met him first at the house of her friend Mrs. Megary in 1860 and he later took up residence at 14 Fifth Avenue. Mary Ann denied that he lived there before they were married, but it is hard to reconcile this with the dates of her movements, since she also said that she left the Fifth Avenue house in April 1862 and moved with four of her daughters to a hotel, prior to moving to the house on Twenty-eighth Street. The rest of the children, including her married son, Gus, remained at Fifth Avenue, together with the servants. Voulettie alone, now Mrs. William Proctor, knew of the trip to Boston, though it is not clear whether at this stage she knew of its purpose. Mary Ann, despite the fact that she now had her freedom (if indeed she had ever relinquished it), felt—rightly, as it turned out—that she would do best to keep this new marriage a secret.

The secret, however, did not last long. In January 1863, when the couple had been married six months, Mary Ann had a bad fall, "and thinking I was not going to recover, I told my daughter of my marriage with Foster." This was again Voulettie, who had always been her mother's confidante. At this time Voulettie and William Proctor were sharing the Twenty-eighth Street house with Mary Ann. Voulettie naturally passed on this hot piece of news to her husband, and it seems that Proctor reported it to his father-in-law, who was also his employer and with whom he was on excellent terms.

All hell now broke loose around poor Mary Ann's head. Both Singer and Clark saw this indiscreet second marriage as a heaven-sent chance to rid themselves of something they would be better without: Singer of an unwanted moral and financial burden, Clark of at least one of the incubi his partner seemed to have brought down around the firm. Mary Ann was summoned to the office of her lawyer, Mr. Fuller, who informed her that in his opinion she had "kicked away the best dish of judicial milk ever set before anyone."

He sent for me to come there and bring my contract—this was the contract in which I compromised my alimony. I had not been to his office since I paid him the $500 for prosecuting my divorce suit. When I went into his office he said, "Have you got the agreement?" and I answered, "Yes"; he took it out of my hand, and after locking it up said, "Mr. Singer is in a dreadful rage and threatens to have you sent to the State Prison." I asked him for what, and he answered, "For marrying again."

Singer was apparently now contending that she *was* married to him, had never been divorced from him, and had therefore committed bigamy in marrying Foster. "He then produced a paper," Mary Ann went on,

> and laying it before me told me to put my name to it. I asked him was it correct, and he said it was. When I went home I found Mr. McKenzie, Singer's agent, in my house, and he told me to pack my wearing apparel and take my children away and we would be back in two or three weeks. I then went back to Fuller's office and he handed me a check for $2,500, which was drawn, as Fuller told me, by Mr. Clark, Singer's lawyer. Next day I called on Mr. Clark and told him that the paper I had signed was fraudulent.

We may be sure that Clark was unmoved by this plea and that he himself had made very sure that the renunciation document was legally completely watertight.

It was obviously either very foolish or very trusting of Mary Ann to sign that paper without reading it first—always assuming she could read; but even if she could not, she could have got Fuller to read it out to her. Perhaps she thought, not unreasonably, that her own lawyer would be sure to be on her side. It was later alleged that Singer had paid Fuller $5,000 to act in *his*, not Mary Ann's, interests. This seems probable: it is hard to think of any other explanation for his behavior. She later sued for the document to be set aside, on the grounds that it had been obtained from her under false pretenses, but was unsuccessful in her suit.

Singer had now acquired yet another romantic interest—an attractive young woman, half French and half English, named Isabella Eugenie Boyer Summerville. Isabella's mother, a lady from Suffolk called Pamela Lockwood who had married a Frenchman, Louis Boyer, ran a pension in Paris, and the story commonly put about by the press after Singer's death was that Singer met Isabella there. He first (ran the story) paid his addresses to Mme. Boyer, but on making

the acquaintance of Isabella transferred his affections from the mother
to the daughter. Mme. Boyer was naturally chagrined, but swallowed
her pride and urged her daughter to go ahead and do as well for herself as she could—it being apparent that Mr. Singer was a considerable
parti and, as he no doubt assured them, unmarried. So, at any rate,
the newspapers said; but this is not the story believed by the family,
who assert that Isabella first met Isaac not in France but in America,
where she was living after her marriage to a Mr. Summerville. Attracted by Isaac's person or his money or (probably) by both, she left Mr.
Summerville for Mr. Singer.

Whichever side of the Atlantic they first met, the pair certainly
spent a considerable time touring around Europe together. Singer
became very attached to Isabella. Traveling back across the Atlantic
early in 1863, they installed themselves in the Fifth Avenue house, now
conveniently vacant. On 13 June 1863, seven weeks after Singer had
obtained the renunciation document from Mary Ann, he married
Isabella, now in an advanced state of pregnancy, in the Church of St.
John the Evangelist, Waverley Place. (Isabella had obtained a divorce
from Mr. Summerville, so that the expected child could be duly
legitimized.) Perhaps if Mary Ann had had her wits about her she
could in her turn have sued for bigamy, but she did not do so.

On the face of it this marriage might seem the most understandable thing in the world. Singer was now approaching his fifty-second
birthday, an age at which even the wildest spirit might feel like
settling down. He was now finally and officially quit of Catharine and
Mary Ann. As for the rest, they had always known where they stood.
Now he was once more enamored of a pretty young woman and about
to become a father for the nineteenth time. What could be more
natural than to marry her?

The truth, however, is not so simple. What had happened was
that Catharine, who had in theory been dealt with once and for all in
1860, was now trying to stage a reappearance. Perhaps she had heard
of the generous settlement made on Mary Ann. At any rate, she was
now alleging that she had been fraudulently persuaded into accepting
her own $10,000 settlement, and that Singer's fortunes at that time
had by no means been "falling" as had been represented to her. She,

therefore, proposed to reopen the case and have her rights. William Singer described how in July 1863 his father came to see him at the Singer factory where he was working, and told him that his mother had once again begun proceedings, but that he (Singer) would beat her, as she had neither money nor energy, while he had both. Singer then revealed to his son that he had married again. William asked him how long he had been married and he said about a month but that Isabella, his wife, was about to become a mother. He said that she had traveled with him as a companion across Europe for about two years and that he had married so as to defeat William's mother, since by bringing an innocent party into the case he would have the sympathy of the court. (This ruse seems to have worked.) William then asked his father how many children he had had by his women and he said about twenty. He told his son that he would want him to "act right" in this matter. A few days later Singer returned and told William that he wanted William to testify for him in the case. William begged his father to let him remain neutral, but Singer told him that, if he would do what he wanted him to do, he would give him a furnished house and make him a rich man. "Take your choice, your mother with poverty or me with riches," he said. William repeated that he wanted to remain neutral, but that if he had to take his choice, he would not perjure himself as to what had happened in 1860. At this Singer became violently angry and said his son was "the wickedest of men and the silliest of fools." He became very abusive and threatened to murder William if he himself "hung as high as Haman the next hour." Then he went. He never forgave his son. For William's part, this strange scene remained vividly in his memory for the rest of his life, which is perhaps scarcely surprising.

9

The Castle and The Wigwam

Mary Ann asserted that when Singer returned from Europe he occupied himself exclusively with "riotous living." But in fact it must have been some years since his personal preoccupations had left him with very much energy to devote to the business. Certainly there can have been little time or attention free for the inventions and improvements which had been his main contribution to the firm's progress throughout the 1850s. His business judgment could be faulty, as the comment on the foreign agencies being a "waste of time" shows; he voiced this opinion in 1861 and by 1863–64 the pound sterling was worth thirteen to fourteen dollars because of the Civil War, so that any machine sold abroad was worth much more to the firm than one sold (when one could be sold) in America. Singer in Europe, preoccupied with pleasure, drained the firm's resources there without noticeably contributing to its running. In America he was not much more help. He grew increasingly absent-minded, tending to put important letters in his pocket and forgetting to draw Clark's attention to them until several days later—a failing which his meticulous partner found most annoying. In addition to all this, as more and more children appeared who could claim Singer as their father, Clark grew increasingly worried about the fate of the company should his partner die suddenly. It looked as if Singer's will—if, indeed, he had troubled to make one—might well be a complex document and would very probably be contested, whatever provisions it might make. How could the company possibly be run in such an eventuality? The litigation which would ensue should Singer die intestate simply did not bear thinking about.

It is not surprising, then, to learn that for some years Edward Clark had found the original partnership arrangement unsatisfactory. Not only was his partner no longer contributing anything substantial to the business but (Clark felt) he was actually becoming a liability to it. It was not merely that Clark personally disapproved of Singer's amorous exploits and resented the publicity they seemed constantly to have been attracting since 1860. They were also very bad for business. We have seen that an important part of the firm's selling platform had been a particular appeal to such respectable members of the community as ministers and their wives; the effect of the recent publicity on this market, even if mercifully dimmed by the war, could be only too easily imagined. Singer, after all, was the partner who had always been publicly identified with the machines which bore his name—indeed, a great publicity point had always been made of his personal involvement with their development and production. This might now prove a serious drawback.

Clark also had other, more personal grounds for dissatisfaction. As the business grew it had become increasingly irksome to him that the partner who got all the publicity was Singer while he, Edward Clark, was contributing just as much, if not more, to its success. The partnership was after all a fifty-fifty one, and Clark felt it should be so in all ways, not just financially. The firm and the machine both bore Singer's name; why then should Clark not at least have the title of president of the company? Singer, however, would never hear of such a thing.

Obviously business could not go on like this indefinitely. The change Clark proposed was the conversion of the company into a joint stock corporation in which each of the partners would hold shares and be a director but which would absolve Singer from the responsibilities of active management, allowing his involvement with the firm to become a purely financial one. Clark thought he had obtained Singer's consent to such an arrangement in 1862, but he was doomed to disappointment. In July of that year Singer wrote from Exeter in England (where Clark probably preferred not to wonder what he was doing):

"When I left New York, I thought that there was a perfect under-standing between us as regards a joint stock company. I have thought it all over and have come to the conclusion that it is not for the benefit of the firm to put it into stock at present . . ." Clark, whose concept of what was or was not beneficial to the firm was quite different, could do nothing but gnash his teeth and await the return of his errant partner. This occurred in 1863, when Singer came back to deal with the dying throes of two previous marriages (one of which might or might not have been a marriage) and contracted a third. Once again Clark made his proposal concerning a joint stock company and this time it was accepted. On 3 August 1863, a notice was published in the daily press:

NOTICE OF DISSOLUTION. The copartnership heretofore existing between the undersigned under the firm name of I. M. Singer & Co., is this day dissolved by mutual consent.—Signed, Isaac M. Singer, Edward Clark, New-York, July 31, 1863.—The business heretofore conducted by Messrs. I. M. Singer & Co., will be continued by "THE SINGER MANUFACTURING COMPANY" at No. 458 Broadway. Inslee A. Hopper, Pres. A. F. Sterling, Sec.

It may be imagined that this arrangement had not been arrived at without a certain amount of wrangling. Singer had only agreed to the end of the partnership under certain conditions, the principal one being that neither of the partners would be president of the new com-pany while the other was alive and that both would "retire from active participation in the management of the business." In other words, Singer was insisting that the arrangement between himself and Clark remain a strictly equal one. If Singer was to become a non-executive director, then so must Clark. If he was so determined to dissociate Singer from the business, this was the price he had to pay. There can be no doubt that Singer was quite aware of how galling Clark would find this arrangement, and that this was a source of considerable pleasure to him. Singer also insisted that he should not be required to contribute future inventions without additional compensation. Financially, the partners agreed to divide between them $40,000 in government bonds belonging to I. M. Singer & Co. which were kept in the vault of the Chemical Bank. On 9 July they signed an agreement to turn over all other assets of the partnership, including $120,000 of New York real estate, in return for stock in the Singer Manufacturing Company to be capitalized at $500,000. Singer and Clark would each hold 40 percent of the company's stock.

There now remained only the problem of who was to be the president of the new company, an office debarred to both natural contenders for it. This problem they resolved in classic style by appointing the office boy, Inslee Hopper—a story subsequently handed down from generation to generation in Wall Street. The accepted version of this tale is that in the summer of 1863 Hopper was the office clerk, earning twenty dollars a week. Singer and Clark were at this time arguing over the exact terms of the new agreement, especially the vexed question of who should head the company. "Maybe I was what in these latter times I hear folks call 'a fresh kid,'" Hopper would recount later.

You see, I was not yet old enough to comprehend what a grand and grave thing is the glory of seeing yourself named president on letter heads and promissory notes. "Take a short cut, gentlemen," I jocularly—which means shamelessly—broke in on their protracted confab late one afternoon. "Be sports—toss a cent—heads or tails!" "Sonny," said Isaac Singer—and ah, with what a pained look,—"sonny, good night." Not at all pathetic echoed Edward Clark's lawyer voice, "Good night, Mr. Hopper." Next morning, Mr. Singer came to my desk the moment he arrived. I could see he was feeling sorry for me, and of course I understood the reason why—my impertinence due to meet its prompt and proper punishment. Yet my tender-hearted senior boss was obviously straining to alleviate the blow a little. Said he, "My boy, are you married?" He evidently couldn't bring himself to putting a man with a family on the street without a little kindly consideration—very, very plain the situation revealed itself to my quickened conscience . . . Still I was young enough to be sure that I was being martyred and I affected the light and airy. "Not married, Sir," I answered. "Are you picking a fifty-fifty bride for me?" "It might be an accommodation," said Mr. Singer. And the dear old boy stood there in a real appealing attitude confiding: "We are going to incorporate the business, and Clark won't let me be president—and

I swear I won't let him." Right then in came partner Clark; and almost they fell into a chorus like this: "Our president ought to be a married man—the office requires some dignity. You are pretty young but we think if you were married we could try to get along with you for president. Don't you know some nice girl you would like to marry?" I did know such a girl, but the size of my salary had never let me hint the matter to her. That evening I hurried round and told her my dilemma. She was nice about it, and five weeks later we were married and I was drawing $25,000 a year salary.

This romantic tale is surprisingly near the truth, though of course embroidered in some details. Hopper had had in fact a good deal of responsibility already in the firm. He joined as a messenger and clerk in 1857 but was soon running the office whenever both partners were absent. By 1860 he was reorganizing the company's office in Richmond, Virginia, visited Havana, Cuba to appoint an agent and took charge of the branch at Newark, New Jersey. In 1861 he was recalled to become cashier of the main office. He did indeed get married just a few weeks before incorporation. But his salary did not reach $25,000 a year until 1875: his starting salary as president was just $10 a week more than he had been making before, increased to $6,000 a year in 1865.

This marks the end of Singer's active participation in the firm, but he by no means relinquished a close interest in it. He was of course a member of the Board of Trustees; also on the board was his trusted friend and agent George Ross McKenzie, and his son-in-law William Proctor. Proctor was vice-president until 1866, when McKenzie took over the job. Singer interests were thus amply represented. One Singer, however, did not do so well under the new regime: Gus Singer was sacked by Hopper for giving insufficient attention to his work in the Newark branch office. It was an indication that things had indeed changed, since nobody would have dared take such action while Isaac was still the senior partner. Other members of the family, however, continued as before—Charles Sponsler, for example, remained in charge of the Baltimore office.

Singer and Clark were, of course, not the only shareholders in the Singer Manufacturing Company, though their holdings were by far the largest. They sold 175 of the company's normal par $100 shares to each of the four officers at $200 a share; these were paid for largely by dividends, personal notes at seven percent interest being held meanwhile. Ownership of this number of shares was a condition of being an officer of the company. Twenty-three other employees were also given the opportunity to purchase shares at the same price: eleven failed to take up the offer and in the end the number of shareholders was nineteen. The eleven non-participants no doubt soon regretted their decision. An initial dividend of twenty dollars a share was paid in October 1863, after only a few months of incorporation. Dividends of fifty dollars were paid on each share in 1867, and eighty-five dollars in 1865. Dividends paid for the shares in less than five years; from then on it was all profit.

★   ★   ★

Singer now found himself excellently situated: both financially and personally, he was better off than ever before in his life. He was assured of a comfortable, indeed, a princely income for which he no longer had to work. At the age of fifty-two he was still in his prime, and there were no business worries to distract him from his pleasures (not that they had ever done so to any noticeable extent). His personal life, for the first time for many years, appeared to be more or less in order. He was living with only one woman, to whom moreover he was legally married. His wife was young and pretty, and the first of a new series of children had already been born: Adam Mortimer Singer, born in New York City in 1863. Furthermore, this was a wife before whom he did not have to pretend to be what he was not. Isabella must have known from the start all about her husband's previous history; if he did not tell her while they were in Europe, he would have had to do so once they arrived in America, when they were faced with both Catharine and Mary Ann on the warpath. Isabella, who was an intelligent girl, cannot have been very shocked and was scarcely in a position to pretend to be so. Her reaction was to make friends, as far as she could, with the various Singer offspring; they must have provided almost the only companionship she could find of her own age.

Voulettie, who like Isabella was in her early twenties, seems to have been particularly kind to her, and she also made friends with John, Voulettie's brother, who was just twenty at this time. Isaac was never lacking in paternal feeling and he seems to have wanted Isabella to mother any of his younger children who might need it. She certainly seems to have played this role with Alice "Merritt," the daughter of Mary Eastwood Walters, and with Caroline, Mary Ann's youngest child. It seems unlikely that she met any of the "Mathews" family at this time, since they were hidden in San Francisco, but she certainly knew of them. What she thought of them all we do not know; the attitudes she expressed in letters concerning them, even after her husband's death, are conventionally dutiful and affectionate. They were probably viewed, since she was always practical, as inevitable adjuncts of her own very advantageous set-up; the last thing she wanted to do was alienate anybody, least of all her husband. It seems probable that both Mary Ann and Gus met her with hostility, since she sometimes mentions them with a certain asperity. This attitude on the part of Mary Ann at least was hardly surprising. But now for the first time Singer was able openly to play the role of proud father to all his considerable tribe of offspring.

The setting for this new life of leisure and respectability, however, was not to be Fifth Avenue. Desirable as the location might be, Isabella may have felt that they should make a tangible break with the old life and with the melodrama which overshadowed No. 14; or perhaps she merely found the climate of New York oppressive and unpleasant, as indeed it can be. At any rate, in the spring of 1864 they moved out of the city to the semi-suburban town of Yonkers on the Hudson River, where the air was healthier. They took with them, as well as the baby, Singer's youngest children by Mary Ann, Caroline and Julia (now aged seven and eight) of whom their father had claimed custody since their mother's unfortunate marriage.

They moved in the first instance to a house at the top of Locust Hill Avenue which had belonged to Judge Anson Baldwin, but this was merely a staging post. Their first real home was to be no ready-built, second-hand rubbish. Singer bought several hundred acres of land on the outskirts of the town and began building what was to be

known as The Castle. This was very much the architectural equivalent of the canary-yellow thirty-one-seater coach and, like all his
constructions, with or without wheels, tended to the capacious rather
than the stylish. It was built entirely of granite and was filled with the
most costly and elegant furniture—though whether Isabella found
elegance in Yonkers quite the same as elegance in Paris is another
question.

The house in Fifth Avenue had been in the fashionable part of
town, but that was not its only attribute. It was also convenient for
business and easy access to passing friends. If the guest list had not
always been distinguished during Mary Ann's reign, there was at
least always a sociable bustle about the place, full of children, friends
and family. The Castle was quite a different proposition. It was
large and isolated and designed for the reception of a large number of
guests, and, for the first time, Singer had leisure to think about their
presence or absence. The New York Singer had been a very busy man,
occupied with his business and his multifarious personal lives. If
society chose to ignore him he would scarcely have noticed, and
would not have been bothered had he done so; he had better things
to do with his time than worry about such trifles, which in no way
interfered with his enjoyment of life nor (if we are to believe her
accounts) with that of Mary Ann. But life in Yonkers was another
matter. The Yonkers Singer was a gentleman of leisure married to a
pretty, young and sociable woman, and to such a ménage social life
is generally important. Furthermore, the way in which the ménage
had been set up indicated very clearly the scale which this social life
was expected to attain. Singer was now a gentleman rich, celebrated
(for a variety of reasons) and willing and able to entertain in a sumptuous manner. He determined, therefore, to start as he meant to go
on and set the ball rolling in the right way.

When the construction of the new house was finished, great
preparations were made for a house-warming party. It was a disaster.
In the words of one newspaper reporter, "Hundreds were invited.
Few went. Singer's previous life had been the topic in nearly every
circle. Residents in Yonkers ignored him, and few of his old associates
clung to him." For the first time, Singer was forced to confront the

social consequences of his scandalous behavior. Until now he had never had to do this and must have felt confirmed in his conviction that irritating social rules were nonsense and of no consequence. For thirteen years he had more or less openly flouted them, and what had been the result? Life went on exactly as he would have wished it to. When things got too hot he simply escaped abroad, leaving the unfortunate Clark to bear the brunt of the social ignominy which by right should have been heaped on his partner. Singer's reaction to this was that if Clark was stupid enough to care about such things, he deserved all he got. But now Singer, too, wished to enter respectable society—even if only that of Yonkers—and now he found that he was unable to have his cake and eat it. For the first time since he had known Singer, Clark must have felt that natural justice was being done.

Such a life was obviously intolerable, for Isabella even more than for Singer. They stuck it out for two years. Two more children were born to them—Winnaretta Eugenie in January 1865 and Washington Merritt Grant in June 1866. Late in 1866 The Castle was closed up and let to a Mr. Waring, a wealthy Yonkers hat manufacturer. "Occasionally, in the cleaning season, the big open carriage that had once excited the imagination of throngs in the Park, was drawn out of its place in the big stable, but beyond connection with that, their names were seldom spoken," recalled the reporter. Julia and Caroline returned to their mother. Defeated, the Singers made their way to a more civilized life in Europe, never again to return to America.

★ ★ ★

Isaac and Isabella's first move in Europe was to Paris—a natural choice since this was Isabella's hometown and her mother and old friends were there. Here, early in 1867, they settled at No. 83 bis Boulevard Malesherbes, a respectable bourgeois address in what Proust called "one of the ugliest districts in Paris." Three more children were born to them: Paris Eugene (named after the town of his birth and not, as he later liked to hint, after the Greek hero) in 1867, Isabelle Blanche in 1869, and Franklin Morse in 1870. Singer, it will be seen, was keeping up his record of alternate boys and girls.

However, they were not fated to settle in Paris either. In August 1870 the Franco-Prussian War broke out and, with Bismarck's armies approaching, the town was obviously no place for foreigners who had no need to remain. The Singers felt disinclined to face it out but it seemed at one point as if they might have no choice: Isabella was pregnant and in no state to travel and the baby did not come. As the German armies drew closer and it still did not arrive, the Singers began to wonder if they would ever get out in time. However, they were eventually able to leave, bearing with them the newborn Franklin, a fortnight before the last train departed from the Channel coast; they there took ship for England. Paris Singer, who was three years old at the time, remembered that the train moved literally at walking pace, as a man went in front to see that the line had not been torn up.

They settled first in London at Brown's Hotel, but Isabella's health was delicate after the birth of Franklin and a doctor recommended the mild sea air of Torquay on the south Devon coast. They accordingly decamped for Devon and, having arrived at Torquay, decided they liked it very much—so much so that this was where they would live. The countryside was beautiful and the little town was lively and cosmopolitan, with visitors of many nationalities. The Singers resided at the Victoria and Albert Hotel in Belgrave Road, but this was not very satisfactory and they were always having to "hush the children down," so they looked around for some more suitable permanent home. This, as might be expected from the scale on which Singer now envisioned "home," was not easy to find. They first thought of buying the Brunel estate in the hills above Watcombe, but were told this was not for sale. They then tried, equally unsuccessfully, to buy some high ground above Torquay. Eventually they settled upon the Fernham estate at Paignton, just west of Torquay. This already contained a house which was called Oldway House; but, as before, the Singers would settle for nothing secondhand. Singer had got to know a young local architect, George Bridgman, who had been designing an extension and improvements for the Victoria and Albert, and he now commissioned Bridgman to build him a house which would be a fitting successor to The Castle. Castles, however, were ten a penny in Britain, and Singer was in no doubt as to the name of his new house. "I want a big wigwam, and I shall name it The Wigwam," he declared. Many years later his son William wrote that "our dear Father, Isaac Merritt Singer, *never for a moment failed to claim and insist* (no matter where he happened to be temporarily) that he was an American citizen and always in fact an American. That is why he named his Paignton home 'The Wigwam,' an Indian name for home."

Patriotism, however, was not to be carried to absurd lengths: The Wigwam was to be a wigwam in name only. In design it was to be, as the architect's daughter remembered, "as far as possible in the style of a rather florid French villa." Its luxury was to be such as to outshine all the surrounding palaces of the local landed gentry. It was to have its own private theater (no doubt an old dream fulfilled at last), riding school and arena, conservatories, gardens (which are still a showpiece) and all the latest and most modern fittings. As many rooms as possible were to face south, as Singer felt the cold badly. He supplemented his instructions with his own sketches and sent Bridgman off to France to make a special note of items he particularly desired.

As a client Singer seems in many ways to have embodied every architect's private nightmare. During the construction he was constantly on the site, and if any part was built that did not please him, even though it might be to his own original design, he would have it pulled down and rebuilt more to his liking. Of the finished product, one observer declared that "it does not possess any marked architectural features, but it is capacious," which was, as the reader will recognize, highly characteristic. The design of the new Oldway House (the original was renamed Little Oldway) suffered only one setback, which was that Singer originally wanted to buy not only that land which was up for sale but also the whole of the adjoining property, so that he might have uninterrupted sea views and own all the frontage along the main Torquay road. He was, however, unable to do this because one lady who owned a house and garden in a key position refused to sell. He did, nevertheless, manage to buy up about twenty cottages, gardens and various other lands, and the whole property was surrounded with a fence.

The observer just quoted was not quite fair in asserting that the new house had no distinguishing architectural features. It did have one (although this was perhaps not strictly part of the house): a great circular covered arena, with a removable floor which could be taken up so that horses could be exercised there and riding lessons given to the children in bad weather, and relaid to be used for parties and performances. (Singer's original intention was to engage companies of artistes to entertain himself and his friends, and at least one circus is said to have performed there.) Since there was already a house on the estate in which the family could live while work was in progress, this arena was built first, together with stables, harness room and carriage accommodation. As soon as this was done, Singer returned to coachbuilding. George Bridgman's daughter Laura remembered that

> as soon as the Arena was in hand Mr. Singer gave Messrs. Edwards, Coachbuilders, Torquay, an order for two large vehicles: a coach for travelling by road as he did not like railway travel, fitted with every convenience for picnicking, and a high coach with four horses to attend the local race meetings. The coaches were exhibited at Messrs. Edwards' premises and were an attraction. I don't think the Race Coach was used many times, but it caused a sensation as it dashed round the New Road and the streets to St Marychurch races.

If the Singer of Paignton days was remembered for his coaches, his parties left an even deeper impression. With his histrionic inclinations and his taste for spectacle, it was obvious that Singer was potentially an epic party-giver. However, this potential had had little opportunity for fulfillment until now. There had, of course, been the sewing machine balls at which, with Mary Ann on one arm and Voulettie on the other, he had cut so magnificent a figure, but socially speaking these had left something to be desired, as had the annual New Year's parties for the employees of I. M. Singer & Co. The opportunity to entertain on a less commercial scale had so far been denied him, but he was determined now to make up for lost time. His parties were superb affairs enjoyed by everyone and most of all, one suspects, by their impresario. "My memories of Mr. Isaac Merritt Singer are very vivid still," wrote Laura Bridgman.

He was most impressive in appearance: a handsome old gentleman of medium height with a white "Father Christmas" square cut beard and, when dressed in his party attire, he was to us children magnificent. He felt the cold very much . . . and usually wore an overcoat, but for a party this overcoat was of velvet-lined satin. I remember a large garden-party (before the Oldway grounds were disturbed for building) of children and grown-ups. He was immaculately dressed in morning attire but with a long coat of royal-blue velvet lined with primrose satin. It had a striking effect but it seemed to suit him and looked quite in order—for him—and on other occasions he wore similar coats in different colours. After the Arena was finished several large parties were held there . . . The guests at these parties were principally children with a good sprinkling of grown-up young people. Entertainments by Conjuror or Christmas Tree [*sic*], as well as the Italian Band from Torquay, was added to the joy and excitement. During the afternoon Mr. Singer would go up to the little gallery in the Arena and give a speech of welcome to his guests. The Italian Band, a musical standby for many years in Torquay, would be playing in the Carriage stand and all of us dancing or playing inside the circle . . .

Despite this fatherly or, indeed, grandfatherly image, other old tastes were also not quite dead. "The parties and balls at 'The Wigwam' were lavish affairs and to these the better-class residents and the wives and daughters of the leading tradesfolk of the district were bidden," recalled another Paignton resident. "There are still in the town some fine pieces of jewellery, family heirlooms now handed down by some belles of the ball, who pleased the millionaire and next day received an intriguing package." It seems, however, that it was by now only the eye which roved.

It will be noted that despite their magnificence, the guest lists— "tradesfolk," "better-class residents"—for these parties were still not very distinguished. The reason for this was that the Singers' first social experience in Torquay was not so very different from that of Yonkers; save that in Torquay this was probably a generic, not a personal snub, for being in patent sewing machines rather than for scandalous notoriety. "He tried to get into society by giving a grand ball, to which all the aristocracy of the neighbourhood were invited," wrote a local newspaper some years later. "But they mercilessly snubbed him, and in revenge he asked all the tradesmen of the place, and gave them an entertainment the like of which for magnificence has hardly ever been seen in England." This could reasonably be construed as the act of a true, democratic American, and no doubt Singer liked to see it that way. It was the kind of snook cocking he had always enjoyed, and what the aristocracy disdained, the ordinary folk certainly appreciated. Singer had three "special days" every year: Christmas Day, when he had several hundredweight of meat and provisions distributed to the poor of the district; the Fourth of July, when, as a patriotic American, he gave a party; and his birthday in October, when he gave treats to children.

It seems unlikely that anyone in Torquay at this time knew much about Singer's previous personal history. The New York papers were not current reading in Devon and the scandal had hardly been such as to reach the European press, which had scandals enough of its own. Indeed, it is still possible to find residents who declare that tales of eighteen other children (apart from the six who were brought up at Oldway) are nothing but a slur and a slander invented by Singer's enemies to discredit his memory. However, those who knew the family well were vaguely aware that this was Singer's second marriage and that he had other children—by, it was assumed, his first wife— in America. Such a situation, among Americans, was not unusual, and anyway, who was to say she had not died, leaving him a widower? The fact of the existence of a first wife seemed indisputable because, after the Arena was finished and when the building of the new house was getting under way (its foundation stone was laid by Isabella on 10 May 1873), two daughters joined the family from America: Alice, very tall and fair, and Caroline, pretty and petite with dark hair and complexion. It would scarcely have occurred to the good burghers of Torquay to attribute their very different appearances to the simple fact that the two girls had two different mothers (Mary Eastwood Walters and Mary Ann)—neither of whom had been actually married to their father.

Among Singer's many children Alice Eastwood Merritt was one of his prime favorites. It is easy to see why, for she seems to have been very like her father in almost every way. She looked like him, being tall and fair and (if her portraits are anything to go by) very pretty, and she shared his predilection for the stage—she later made quite a name for herself as an actress under the name of Agnes Leonard. In 1873 she would have been twenty-one or twenty-two, and had until then been living with her mother, "Mrs. Merritt," in fairly poor circumstances in Brooklyn. She was a bright girl and made a little money giving readings and recitations, describing herself as an "elocutionist"—just as her father had done, thirty years before.

Singer could no longer abide "Mrs. Merritt" but this in no way affected his feelings for her daughter. Once the family was well established at Oldway therefore, he invited Alice to come and join them, where she could enjoy a much more luxurious life and better social opportunities than in Brooklyn. (The fact that the house had its own private theater was perhaps a further temptation.) Alice at first refused his invitation, preferring to stay with her mother, but she eventually relented and was in England in time to take part in Singer's sixty-second birthday celebrations on 27 October 1873. These included an entertainment put on by the children and their friends: the ritual bow to Shakespeare, a scene from *Henry VIII*, "Cardinal Wolsey and Cromwell," given by Mortimer; two solos, "Spooning in the Sands" and "Courting in the Rain," by Winnaretta and Mortimer; and, as a climax, "Breaking the Spell," a comic opera in one act in which the part of Jenny Wood, maid of the Rising Sun Tavern, was played by Alice (now known as Miss Alice M. Singer). Other parts in the opera were taken by a Mr. Rice, who was obviously a good friend since he participated in similar entertainments given in later years, and Mr. John Cushway, Isabella's cousin's husband, who acted as Singer's confidential secretary. The pianoforte was played by Miss Florence Matcham, the daughter of the building contractor for Oldway. Also living with them at this time was Isabella's sister, Mme. Boyer (who later became Lady Synge). This gives a fair idea of the composition of the intimate family parties at Oldway; they might not have been quite socially *comme il faut*, but it sounds as if they were fun.

Alice seems to have inherited yet another of her father's characteristics: she, too, was to have a flighty and well-publicized amorous career, the first stages of which were played out at Oldway. No doubt her mother, living in bleak respectability in Brooklyn and subsisting largely on $15 a week allowed her by Singer, felt able to ascribe her daughter's subsequent troubles to the pernicious influence of her father.

Among the habitués of Oldway at this time was a Monsieur (or possibly Mr.) La Grove. He was generally known as "the handsome Frenchman," and may have been a friend of Isabella's from Paris, though some sources imply that he was in fact American (and his subsequent behavior would seem to support this view). Certainly his first names do not sound very French: he was called William Alonso Paul La Grove.★

La Grove and Alice fell in love with each other, no doubt much to Singer's gratification; here was proof that Alice had done the right thing in coming to join her father in England. Their wedding was planned for June 1875. It was to be a magnificent affair: the bride's outfit, of heavy white satin and Brussels lace, cost more than £2,000, her trousseau was proportionately splendid, and rumor had it that her marriage portion was more than £200,000, plus about £100,000 in jewels. This was probably an exaggeration, but she certainly was given some jewelry as a wedding present from her father—notably a pair of diamond earrings worth more than £2,000.

★ All these details were sedulously recounted by *Truth* magazine on the occasion of the breakup of Alice's second marriage, to the famous actor Frank Bangs, ten years later. Alice had become infatuated with Bangs, pursued him to the stage door after a performance and persuaded him to come home with her and meet her mother; he stayed at the house for several weeks, and in a couple more, they were married. "On their bridal night," recounted *Truth* with evident relish "the reason for their present state of affairs was given. I cannot speak plainly on the subject but will adopt the language of one intimate friend of Mr. Bangs, who said the real cause of the separation was his refusal to permit the bride to give certain romantic Parisian touches to the crescent joys of the honeymoon . . ." Alice had made a sad mistake: in fact Bangs was very effeminate and was known as "Lillie" Bangs by his colleagues. Alice terrified him, and he fled.

The wedding, however, had to be postponed on account of the illness of the bride's father. Singer had for some time had a weak heart and in May 1875 he caught a chill and was said to be suffering from "an affection of the heart and inflammation of the wind-pipe." He took the very best medical advice—Sir William Jenner came down from London to attend him—but he did not improve. As the bride and groom felt disinclined to wait indefinitely, and as all arrangements had been made, the wedding took place in spite of this illness on 14 July; in the event, Alice was given away by Mr. David Hawley, her father's trusted American lawyer and a close family friend. (They had originally met Hawley in Yonkers; he was one of their few friends in those unhappy days.) Various guests, including George McKenzie and Inslee Hopper, came over from America to attend the celebrations. Singer himself had arranged the festivities—this was a party to outdo all his previous parties—and most of Paignton and Torquay were invited.

Alice's first marriage, like her second, was a short-lived disaster. It lasted little more than a week, when La Grove took ship for America where he became a "commercial gentleman," apparently finding this prospect preferable to living in comfort with Alice. Whatever can she have done, one wonders, to have had such an immediate and galvanic effect on both her husbands despite her apparent advantages, both physical and financial?

What her father's reaction to this news would have been we shall never know, for he died only nine days after the wedding, on 23 July 1875, without ever seeing his great house finally finished. The last party had obviously been too much for him in his weakened state.

★ ★ ★

Characteristically, Singer had not neglected to leave detailed instructions as to the staging of his final role. Mr. Oliver of Union Street, the undertaker, can rarely have supervised such a satisfactory funeral. The body was dressed "in the American style," in a black coat and trousers, white satin waistcoat, white gloves and black slippers. It was placed in three coffins. The first was of cedarwood, lined with white satin and with a white satin mattress and pillow, trimmed with

Maltese lace. The second was of lead "of unusual thickness," and the third was of English oak, grained and paneled, with silver handles and mountings. Singer had left instructions that when he died his own horses, twelve in number and three abreast, should draw the funeral car, of which the sides should be removed so that the gorgeous outer coffin should be visible. These instructions were followed, no doubt to the satisfaction of the two thousand people who attended the funeral. There were four mourning coaches, containing the family and intimate friends.

After the mourning coaches [reported the local paper] came Mr Singer's carriage, with blinds drawn, and drawn by two splendid bay horses. This was followed by carriages containing the men and women servants employed at Oldway, and behind them walked four deep, about 50 men who worked on the estate. Large numbers of private and tradesmen's carriages closed the procession. The funeral slowly wended its way towards Torquay, along the road to which a large number of persons had assembled. By the time the cortege reached the Strand the private carriages had nearly doubled in number; all along the route men, women and children of all stations in life lined the footpaths. The principal establishments at Torquay and Paignton were partially closed, flags lowered to half mast, and at the latter town the church bell was tolled. The funeral procession was nearly three-quarters of a mile long, and contained between seventy and eighty carriages.

It is quite impossible to imagine a private funeral on such a scale today, and it was rather exceptional even then. The local lord of the manor might elicit such a show, but this was a role which Singer had filled for less than four years. During that time he had indeed endeared himself to the local populace—but even more so to the local tradesmen, and it was perhaps they who were among his sincerest mourners. "It need scarcely be said," commented the local paper,

that a man with such abundant means, and with a desire to spend them in reason, promoted the trade of the place in which he resided. Torquay also had a share of his favours for jewels, carriages, costly furniture and apparel. Internal decorations for the house, to the extent of some score of thousands of pounds, have been purchased of the tradesmen in Torquay. Towards the restoration of the parish church of Torquay he gave £100, and the clock fund £80. There were many schemes which he had projected for the benefit of Paignton: it is said that he contemplated the building of a chapel and some school houses. Before he could complete his new home, however, he has been removed, to the sincere grief of the inhabitants of Paignton, to whom he gave promise of opening up a new and prosperous era.

The family had wanted to have Singer buried at Paignton cemetery, but the planned mausoleum took up so much space that the request had to be refused; the mausoleum was therefore erected at Torquay cemetery. One feels that, had he only been able to see it, little in his life would have pleased Singer more than the manner in which he left it. He had without doubt staged one of the best funerals of his day.

10

"A Very Ghastly Domestic Story"

When some personage, however notorious, dies, the brief flurry of obituaries generally marks the end of his public life, so far as the newspapers are concerned. He may re-enter the arena by way of memoirs and biographies, but a decent interval is generally allowed to elapse between death and publication, while the feelings of those who are still living and who may figure in such works are generally spared. Even where this is not true, the fact of their transmogrification into "literature" generally bars them from the pages of the more popular dailies.

Singer was the sort of man whose death is always sincerely regretted by the publishers of mass-circulation newspapers, but in this case they need not have worried. The feast was not yet ended; indeed, the best part—from their point of view—was yet to come. The unfortunate Clark, who had had so much to endure while his partner lived and who had joined the other directors in "sincerely deploring the loss of this distinguished inventor" when he died, was to find that Singer's private life had not died with him. The only consolation was that he was undeniably dead. Clark, as the survivor, was now free to become president of the company—which he did, displacing Hopper, in September 1876.

It all began quietly enough. There were, of course, obituaries, but these skated tactfully over the more tasteless aspects of the subject's life. "As poverty and business difficulties disappeared, family troubles came upon him. He went abroad . . ." murmured the *New*

York Times, while the *Tribune* merely remarked that "the profits on his invention soon made Mr. Singer a wealthy man, and, leaving the country, he took up his residence some time ago in Paris . . ."

Matters, however, were not to be allowed to rest there. Edward Clark had always feared that the extreme complexity of Singer's personal affairs would lead to trouble when it came to dividing up the spoils on his death. His ardent wish that the company itself should not be involved in such quarrels had been one of his main reasons for wanting to change the original partnership into a joint stock company, so that even if ownership of the company's shares was a matter of dispute, its direction was not. Such forebodings were now shown to have been amply justified. Singer had kept successfully out of the headlines for the previous twelve years, during which time he had led a respectable, if not quiet, life with Isabella, to whom he was legally married. This was unprecedented, and no doubt Clark told himself that he had known all along it was too good to last.

Singer had, it turned out, left an estate amounting to between $13 and $15 million (a gigantic sum, worth about ten times that in terms of today's money) and had disposed of this (or most of it, there being some apparently unaccounted for) in two wills. The first of these had been drawn up in Paris on 16 July 1870 (perhaps prompted by the approaching specter of Bismarck) and it dealt with his holdings in America, amounting to nearly $8 million. Most of this consisted of Singer Company stock, of which the testator had owned forty percent. For ease of distribution, Singer, after some small bequests, had decided to divide his estate into sixty equal parts, each part being worth $10,043.18 cash, and $121,975 in stock. These he distributed among the members of his various families. Isabella was to receive four parts; Adam Mortimer, six; Winnaretta, five; Washington, six; Paris, six; Isabelle Blanche, five; Franklin, six. Then,

I give to each of the following named persons whom in this, my will, I call by the surname of Singer; they being my children, born of Mary McGonigal, of San Francisco, who is commonly called by the surname of Mathews:
to Ruth Merritt, two parts;
to Clara, two parts;

to Florence Adelaide, two parts;
to Margaret Alexandria, two parts;
to Charles Alexander, two parts;
To Alice, born of Mary Eastwood Walters, of New York, commonly
called Merritt, two parts.
To each of the following persons whom I call Singer, born of her
who is now Mary Ann Foster, of New York:
to Isaac Augustus, two parts;
to Violetta Theresa [*sic*], nothing;
to John Albert, two parts;
to Fanny Elizabeth, one part;
to Joseph Emmet [*sic*], one part;
to Mary Olive, one part;
to Julia Ann, one part; and
to Caroline Virginia, two parts.

Singer went on to explain that "I do not give my daughter Violetta
Theresa . . . being now the wife of William F. Proctor, any portion of
my said estate, for the reason that through my appointment her said
husband obtained his situation and interest in the Singer Manufacturing Company, and has thus acquired a fortune which places my
daughter above the necessity of any assistance from me."

It will be noticed that neither William nor Lillian, his eldest and
first legitimate children, are mentioned in the list. To William, Singer
left $500 outright, and to Lillian, $10,000; he had neither forgotten
nor forgiven their support of their mother against him in her counter-
suit of 1863. Mary Ann received nothing. It emerged that, besides the
$15 a week he had been paying Mary Eastwood Walters in Brooklyn,
he had been paying Mary McGonigal in San Francisco $1,500 a
quarter for the upkeep of herself and her five children. When Singer
died she was allotted $5,000 a year for herself, but she would neither
accept nor use this money and could scarcely bring herself to go and
live in a house bought for her by her now wealthy children, seeing
that the money had all come from their father who had so disgraced
her and dragged her name through the mud.

In 1873, Singer had made a separate settlement of the English
portion of the estate, which was to go to Isabella and her family. On

14 July 1875 (the day everybody was gathered at the house for Alice's wedding) he had made yet another will, combining that of 1870 with the 1873 settlement, so that all the estates were disposed of in one document; the division was the same as that already stated. Isabella was in all cases residuary legatee.

It is not difficult to deduce from these arrangements where the trouble was likely to come from. William and Lillian were not difficult to deal with. All the illegitimate legatees agreed to contribute toward a settlement for them: they gave $10,000 each, although some, such as Gus, did so very unwillingly. In the end William and Lillian got $45,000 each, and by 1 December 1875, Isabella was able to write to David Hawley, one of the executors, "I expect William, Lillian and their mother are settled with by this time, at least I hope so; for that would be, so much done." Isabella was anxious for the will to be settled since, until it was, none of the money was forthcoming, except by courtesy of the executors, and there were considerable expenses to be met, including the finishing of Oldway. On 7 October 1875, she wrote to Hawley and his co-executor, Dr. Pridham: "Gentlemen—I request that you will with due speed complete the works now in progress on the 'Wigwam' and grounds, including the roads, walls, etc . . ." and the next day, "I do hope affairs are being settled in New York and that we shall not have to quarrel . . ." Such hopes, however, were to be confounded. William and Lillian might have agreed to settle, but Mary Ann—the conspicuous absentee from the lists of bequests—had not, and Mary Ann was absolutely determined to contest the will. Her grounds were simple. She claimed that Singer had never been legally married to Isabella, since he had never legally divorced Mary Ann, and that she, not Isabella, was therefore entitled to the wife's share of the estate. She was quite prepared to admit that her own marriage to Foster had been illegal if she could thereby prove that the same was true of Isabella's marriage to Singer.

Almost everyone concerned had hoped it might be possible to keep the case from actually coming before the court. For one thing, nobody wanted to waste good money on exorbitant lawyers' fees. Mary Ann's children did not particularly relish the prospect of their father's history being once again recited in public; they had been well provided for, and with their incorrigible father out of the way they stood a good chance of making their way in polite society. They therefore proposed to provide for their mother by allowing her the interest on $75,000—a reasonable offer, since she would have nobody but herself to support, she and Foster having had no children. There were also rumors that Isabella's side had offered Mary Ann a generous compromise: they were reportedly willing to pay her $200,000 out of the estate, but "the intimation has had only the effect of encouraging her in her litigious determination to secure about $4,000,000 which she, as the alleged widow of the testator, claims to be her rightful due." All blandishments were scornfully refused by Mary Ann. "What a fine old time you must be having with that amiable tribe; I long to hear the last of it, they certainly have acted a two-faced part with me," wrote Isabella to Hawley on 1 November, by which time it had become quite clear that Mary Ann would take the case to court. Indeed, the proceedings had begun at the end of October; they took place in the surrogate's court, which deals with questions of wills and probate, at White Plains, Westchester County, where Mary Ann was now living. The surrogate in this case was appropriately named Coffin.

The lineup of the two sides was fairly predictable. The Singer Manufacturing Company directors all supported Isabella; as far as they were concerned she was the official widow and they were on her side. They were, above all, anxious to settle out of court if possible, and so avoid the scandal of a court case; they therefore did their level best to get Mary Ann to agree to Isabella's proposals. "Please tell Mr. Mackenzie and Mr. Clark I can never repay them for their kind intercession on my behalf, but I shall ever be grateful to them for what they are kindly doing for me," she wrote to Hawley on 1 November. Several of Mary Ann's children, though not the stubborn Gus, also took Isabella's side—possibly because they wanted to avoid scandal, but possibly also because they felt Isabella was in the right: that, at least, was how she chose to take it. "I must say I feel ever so much happier since I know Mr. and Mrs. Proctor are sincere and that dear Voulettie has not forgotten our old friendship," she wrote on 1 December.

It was clear from the start that the case would fulfill everybody's worst (or best) expectations. The *Herald* noted gleefully that

> to scandal-loving ears, the case, as already foreshadowed, presents an unusually rich treat, for, without desiring to anticipate, there is good authority for the assertion that the litigation commenced is likely to elicit the most astounding domestic revelations which have as yet been spread before the present generation. The testator was a man of noted laxity of morals. Under the terms of the will some twenty or more of his offspring by diverse and sundry mothers are entitled to an average of about $200,000 each. There have been and are other children of the deceased whose names are not to be found in the testamentary document, and it would not be surprising if, before the present legal controversy be closed, his "sons should come from afar and his daughters from the ends of the earth" to be present at and partake of the final distribution of his immense estate. Notable among the spectators in court during the preliminary hearing of the case was a young man who had apparently seen little more than twenty summers. His well tanned face and brawny hands bespoke familiarity with laborious outdoor employment. He was observed to pay the closest attention to the proceedings, and particularly to the arguments advanced by contestants' counsel during the desultory debate. He was evidently a total stranger, not speaking to or being addressed by any one in the court room. After the adjournment, and when on the way to the railroad depot, he approached the writer with an air of familiarity, and on being promptly interrogated as to his identity, merely replied that he had traveled night and day from Kansas in order to be present at the proceedings, as he had an interest in the estate. Having been asked as to what he based his claim on, this mysterious young man said, in a half whisper, "they will find out who I am soon enough," and then bounded across the street.

Alas, he also bounded out of the story, and his identity was never revealed, any more than was that of any other of the unnamed children hinted at by this writer.

Mary Ann, being the only member of the principal cast regularly to appear in court, was, of course, the cynosure of attention. The impression she made was generally favorable. All agreed that it was still possible to see in her traces of that beauty which had once proved so irresistible to the young Singer.

The very first reports of the case attributed to Singer a rather more regular marital history than he had actually had: they represented him as having divorced Catharine in 1840, whereupon he married Mary Ann, whom he subsequently divorced, ending up with Isabella. It soon became clear, however, that this was a considerably idealized version of affairs, representing what should have happened rather than what actually had happened; and the papers naturally began to make the most of the various revelations as they were unrolled. By 1 December, Isabella was seriously worried. "I should feel dreadfully if I thought my marriage would be proved illegal even though quite innocent, for you know the people in England are so particular about such things," she wrote Hawley, exhibiting a truly French pragmatism in her own attitudes toward marriage.

> There has been such scandalous reports in the papers here; I think Gus must send the American papers here to someone, for they get the news in the Devonshire papers as soon as I do from John and are not satisfied to copy but add all they can imagine to it . . . I am glad to know our Counsil [*sic*] are so satisfied with the progress of the case and I do hope the court won't make Foster Mr. Singer's widow, for she ain't his Widow that's certain, and I should think her evidence has proven that plainly enough . . .

Mary Ann was hoping that if she could bring enough witnesses to testify that they had known her and Singer as man and wife, and if she could prove that she had been generally known and accepted as Mrs. Singer during the twenty-three years they had been together, then her claim that she had actually been his wife in common law, even if never actually married, would be accepted by the court as valid. To this end the most intimate details of their life together were revealed before the packed courtroom, down to the fact that "Mr. John Singer, the contestant's son, recollected that his father, in introducing visitors to his mother, spoke of her as his wife, and that the contestant spoke of Mr. Singer as her husband. The witness was required to brush his father's clothes every morning before Mr. and

Mrs. Singer were up, and he was thus enabled to satisfy himself that his father and mother occupied the same room." Mary Ann, who conducted herself with quiet dignity throughout, obviously felt she had nothing to lose by such revelations and millions to gain; but almost everyone else connected with the family was mortified.

The case was a particularly juicy one for the lawyers, who found that it presented them with unusual opportunities for the public display of their moral faculties and the exercise of their oratorical powers. Ex-Judge Porter, for the defendant, opined that

> the rights in question were those which, by the common judgment of men, rose above the consideration of mere property. Isaac M. Singer's life was full of strange vicissitudes, and it would seem that the incidents of his career were to give colour to events that followed his demise . . . Few men ever left so many children as did Singer. He never covered his acts with the cloak of hypocrisy, but recognized in his will all just claims upon him . . . Whether the will was duly executed was a question which did not yet arise. The contest related to the status of a person not recognized in the will as holding any legitimate relations to Mr. Singer. Here was presented for the first time the spectacle of a mother coming into court to assert that she was a bigamist, and was living with a man who was not her husband, for the sake of intercepting one and a half millions given to her own illegitimate children . . .

Mr. Van Pelt, for Mary Ann, indignantly refuted Porter's insinuations. He presented a heartrending picture of his client. "For twelve years her neck had been pressed to earth by the iron heel of her persecutor," he perorated,

> and she has had to suffer in silence lest the threat of prosecution for bigamy should be carried out and entail upon her a still worse fate. But she possessed a clear conscience, a conscience void of offense towards God and man, and in her darkest hours she could trust in Him who said, "A bruised rod shall he not break, and smoking flax shall he not quench." But her merciless persecutor had gone, and even now I can, in imagination, hear him crying out in agony from his place of torment for that mercy which he denied her when he was on earth . . . The real antagonist is Isabella Eugenia Singer,

named in the will as being the wife, but who is, in reality, the *par-*
ticeps criminis in this whole nefarious business. She it was who sup-
planted the claimant in what was left of the miserable heart of her
husband. She sits arrayed like a queen in her palatial residence,
surrounded by almost regal splendor, and in a foreign land,
whither she and the testator fled, unable to endure the ignominy
which followed them in his own country . . .

After a month in which the details of Singer's personal life were
paraded agonizingly through the court and thus, via the daily press,
through all the drawing rooms and barrooms of New York and,
eventually, Devonshire, Surrogate Coffin finally arrived at his decision.
He found that Mary Ann's claim to be Mrs. Singer had from the start
been "undoubtedly without foundation" and declared Isabella to be
the legal widow. This was the signal for an orgy of moralizing on the
part of the papers who, now that the case was over, felt entitled to
give their own opinions on the matter. "Surrogate Coffin's decision,
given at White Plains yesterday, removes the Singer will case from
the public view, and with it a very ghastly domestic story," noted the
Herald with a certain regret. "The second wife's claims have been
negatived, and Isaac M. Singer's will stands as the monument of a
frightfully misshapen life, which will not fail to be studied by the
ethical historian when he endeavors to concentrate in a sentence his
scorn of the picture formed by a gross nature rioting in sudden
wealth." The *Herald* was of the opinion that "there are loathsome
dregs in the ferocity of self-indulgence" and that "fat legacies will not
cover these dregs." Singer's heirs might understandably have dis-
agreed.

The *Daily Graphic* took a quite different view of the affair. "The
striking thing about the will is that he acknowledges all his illegitimate children, calls them all by name, and makes provision for all of
them," it pointed out.

There is no shirking of responsibility, no mealy-mouthed subterfuge, no polite prevarication, but an honest confession of relations
which most men would have shrunk from acknowledging before
folks. Yet this modern polygamist was not only a successful inventor and businessman but a churchgoer and exemplary Christian

so far as the world knew. This is modern civilization. But, however such polygamous practices may be condemned, it is especially creditable to the deceased that he treated his numerous progeny fairly and showed no disposition to shirk his patriarchal responsibilities at the last. There are a good many men who might profit by his example in this respect.

The best that was generally said of Singer's moral habits was more along the lines adopted by ex-Judge Porter, who relegated active religion back a generation: "Singer was a man of very loose and immoral habits—they were bad and yet they were good in other respects—he was not a religious man, with his fast life, nor did he practice hypocrisy. He was led into these habits in early life by bad associations, although his parents were honest and religious people . . ."

Mary Ann was not yet disposed to give up. She found a new supporter, Commodore Cornelius K. Garrison, who "having thoroughly investigated the case of Mrs. Foster, is firmly convinced that her claim to be the widow of Singer is well founded, and that great wrong and injustice has been done to her. He has resolved to stand by her, and to furnish her with the necessary material and to prosecute her case in every possible form . . . In future, therefore, the millions will not be all on one side; but it will be millions against millions . . ." Mary Ann went ahead with the appeal and disregarded her children's wishes that she should give up at this stage: but the appeal judge, Justice Gilbert, upheld the original decision, feeling that, should he not do so, the married state as it was understood throughout the Western world would be dangerously undermined. "A concubine cannot acquire the rights of a wife by survivorship," he declared.

At last, everyone was free to receive their allotted share of wealth. "Alice is with me waiting for her ship to arrive with the fortune from America," wrote Isabella; perhaps Mr. La Grove was now regretting the haste with which he had decamped. Eager to rid themselves of as many undesirable Singer by-blows as possible, the Singer Manufacturing Company, recognizing and catering to the natural human preference for ready cash over pieces of paper, kindly offered to buy the Singer stock of any heirs who might feel disposed to sell. Many took advantage of this offer, a decision which most of them regretted before long. "John seems to have sold out his share cheaply, do you not think so? but I suppose he got ready money," wrote Isabella to Hawley in February.

She continued to take a mild interest in her late husband's various relations, writing, for example, in December 1877:

> I received the other day a letter from John V. Singer's wife Brigit [I. M. Singer's brother and sister-in-law] stating that he had died on November 12, 1877—it was the best thing that could happen for he was such a dreadful sufferer the last few months of his life —so she tells me. I suppose the Sewing Machine Company have stopped the payments I used to make to him since they can have no receipt. Please mention the fact to the company and I should like Brigit Singer to be given as a present from me two hundred dollars to help her until she can get some mode of helping herself.

Next January she wrote: "I note what you say with regard to paying the two hundred dollars to Brigit Singer and thank you for kindly giving the order and I hope it will be the last I shall hear of that party." This seems to be the last appearance of that John Singer who had surfaced spasmodically throughout his brother's life. We hear no more of any Singer brothers and sisters.

I I

The Immigrant's Dream Fulfilled?

Strictly speaking, that was the end of Isaac Merritt Singer; but it is not really the end of the story. Some people's lives are satisfactorily discrete, ending acceptably with their deaths. But the interest we feel in a man like Singer lies as much in what he represented, in the currents of his time which he embodied, as in his actual doings, and those aspects of Singer were, at his death, still incomplete. It was only in the last ten years of his life, for example, that he had so much as begun to ask himself the question of what, now that he had made all that money, he had made it *for*? Now that at last and indisputably he had the dimes, what was he going to do with them? How—if at all —would this wealth benefit his children? How would it affect their prospects in life? How would their lives differ from his own? How did he stand in relation to them? What was his new status—if he had achieved any new status other than that of being Isaac M. Singer, but rich? He had not, by the time he died, really answered any of these questions, and the division of his fortune at his death did little to improve this situation other than by indicating in the most formal possible way that he did indeed feel responsible for all his children regardless of legitimacy; the cases of William and Lillian proving, on the contrary, that legitimacy was for him a perfectly meaningless concept which bore no relation to paternal affection. (Perhaps we can still see here the traces of his unsatisfactory relationship with his own father, though Isaac was undoubtedly legitimate.)

Singer had not been by nature reflective. He had enjoyed what

the day had to offer and had not concerned himself with the consequences, having an apparently perfect faith in life's ability to sort
itself out in the long run. But until his marriage with Isabella, he
might reasonably have argued that there had simply not been time to
do anything more than keep up with the present. After 1863, however, that had no longer been true. Time, after that date, was just
what he did have, and he had occupied himself during the years
between 1864 and 1875 with trying to identify a niche in society
where this time might most suitably be spent. He never completed
his quest. He had identified various areas where he and his family
were *not* acceptable: notably, respectable society in America and high
society in England. Twenty years earlier such rebuffs would have
been supremely unimportant to him. But having achieved wealth and
leisure it was no longer so easy to ignore them. If you have no position
to keep up, you may exist in a social limbo. After 1863 he was no
longer in that happy situation; and, of course, aspirations change.

Let us return to the scene with which this book began: the
immigrant family staring wide-eyed at I. M. Singer's splendiferous
canary-yellow coach. From their point of view there can be no question that the possession of such a thing is in itself a justifiable goal.
To live in the kind of luxury it represented, to be able to drive
around the park in such style—who could ask for more? But the
owner of the coach asks different questions, especially once the first
thrill of possession has worn away. Where, for instance, is he going to
drive in his fine coach? One cannot drive around the park forever, and
the satisfaction of impressing old friends inevitably wanes. When
Adam Singer left Frankfurt he was perhaps not without his own
dreams of striking it rich in the New World. How, in his situation,
would such dreams culminate? It seems inevitable that there would
have been some element of the "I'll show them" variety. "I may seem
nothing now, but wait till I've made it . . . and *then* I'll come back and
they'll see." But when, in the event, his son does go back to Europe,
having apparently fulfilled the immigrant's dream and made a name
and a fortune on the other side of the Atlantic, nothing could be
further from his mind than returning to Frankfurt. What reason
could he possibly have for going to such a backwater? He has no wish

to be reminded of what he has left behind; the important thing is what lies ahead.

For the children who have known nothing but canary-yellow coaches and dream castles, such questions take on a still different aspect. Their point of comparison is with families of equal wealth, little Vanderbilts and Clarks or (in Europe) little lords and ladies. For these grandchildren, their antecedents are not only of no interest, they are probably a positive disadvantage, something to be quickly glossed over. For them only the future must be allowed to exist, at least until they are accepted. The capitalist romance, the immigrant's dream, is not complete, despite all appearances, until such acceptance is achieved.

At the time Isaac Singer died, it will be seen that he had completed only the first two-thirds of this cycle. It was, therefore, up to Isabella and her family to complete the story if they could, to make some social sense of it. If they could not, then what could be expected of Singer's other families? Isabella's children not only had much more money than any of the others, they also possessed the incomparable advantages of both legitimacy and a settled home background. All these things were extremely important—and particularly so for the boys. Boys and girls, during this time, posed quite different problems from the point of view of advantageous social disposal. Girls, unless they were quite exceptional, took on the social standing of their husband. For a pretty girl with a large personal fortune, the quality of the social standing tended to bear a relation to the size of the fortune; the problem was thus a straightforward one. Boys, on the other hand, had to create their own position in life. A fortune might be an asset, but only if accompanied by the ability to handle it. This required an altogether stronger character, reinforced by firm handling in youth. These were attributes which none of Singer's sons by his first four liaisons seemed to possess.

Indeed, the fate of the boys by comparison with the girls seems either to indicate that the girls were generally much steadier and stronger characters or else to bear out the contention that the social lot of a moderately rich girl was easier than that of an equivalently rich young man. Charles Mathews, for example, Mary McGonigal's son, became a stockbroker. He was also unfortunately an incessant gambler, got hopelessly into debt and threw himself out of a window while still in his twenties. His sisters, by comparison, managed to marry into the English upper classes, where they led satisfactorily rich and respectable lives. Of the whole brood, Voulettie was the only one to achieve total social acceptance in America, where she led an irreproachable life and founded an irreproachable family. She had, of course, the good fortune to marry William Proctor, or at least be irretrievably engaged to him, before the worst of the scandal broke.

The experience of Isabella's sons seemed to indicate that male heirs, if they were to survive, needed considerably deeper purses than their sisters. Her four sons, in fact, all grew up to preserve and increase their fortunes, but the indications are that these fortunes only survived their spendthrift youth because the amounts were so absolutely enormous, and only just survived despite the most careful advice and handling on the part of assorted advisers, executors, guardians and lawyers. As for her own daughters, the fiasco of poor Alice's marriage must have confirmed Isabella in her ideas of exactly what must not happen to them. It was obvious that money alone would not be enough to ensure them the kind of future she considered suitable and which would provide a fitting destination for the Singer fortune. When it came to the crucial question of finding a husband, Alice suffered absolutely no financial disadvantages. She had been one of her father's favorite children, and the first to be married from his splendid new mansion. No expense had been spared, and Alice had had the most magnificent wedding her father's flamboyant soul could devise. What had it got her? A nothing; a potential "commercial gentleman" and an inconstant one at that. The magnificence of the bride had in no way affected the essential insignificance of the bride-groom—on the contrary it was she who was diminished. Of course, poor Alice had had no mama on the spot to cast a critical eye over her intended. Isabella's daughters did, and they must at all costs do better than that.

The problem was to provide the children with a suitable launch-ing pad. For the moment, Isabella chose to remain at The Wigwam, where her sons could be provided with the English education which was considered so desirable all over the world. Armed with this and with their vast wealth and expectations they should encounter few difficulties. Daughters were educated by governesses at home, and such education could as easily take place at The Wigwam as any-where—for the moment. For the next few years, then, Isabella remained in Paignton while the boys were sent off to school as they became old enough—though none of them showed much perseverance with education; Mortimer and Paris were both sent to Cambridge, but neither completed a degree course.

But this life, though pleasant in many ways, could not go on for-ever. The girls were growing up, and Paignton as a social center had its limits. Miss Florence Matcham might be a nice girl and a pleasant friend, but the Misses Singer would have to do better when childhood merged into real life. London, of course, was a different matter, but Isabella knew nobody in London, while the local county families who did and who might have been able to effect some introductions had made it clear that the Singers did not qualify for entry into their society. Paris, however, offered few of these difficulties. This was where Isabella had grown up. She had her roots there; it was where she and Singer had chosen to live when they first came to Europe; and she knew how to set about social life there as she did not in London. In 1879, therefore, Isabella shifted her center of operations.

The first necessity was to make a suitable mark in the correct circles for herself, and this for a good-looking woman still in her thirties, endowed now with immense wealth, was unlikely to present too many difficulties. She was rid of the uncouth presence of her husband, who had left her with only his most socially desirable at-tributes: his fortune and (if she played her cards right) his children. She was, from the beginning of her career in Paris, a well-known figure. As the good-looking French widow of an American industrialist, she was called upon to be Bartholdi's model for the Statue of Liberty, then being prepared for presentation by the people of France to the Americans on the occasion of the centennial of the Republic (although it arrived several years too late for the actual celebrations). She was much courted, and it was apparent that she would not remain a widow for long; this would only have been a waste of such unusual resources, and besides, as a married woman she could do much more for her daughters.

When she finally decided upon a husband, however, her choice was not generally considered a very happy one. She decided to marry one Victor Reubsaet, a Belgian musician who had been (it was rumored) a close friend of hers before she had met Mr. Singer. He was fair game for the gossip columnists, who called him a "tradesman" and who sneered at his inheritance, from an uncle, of the title Vicomte d'Estemburgh (sometimes spelled "des Tambourgs"). Isabella's delight at this fortuitous circumstance must have been doubled when another of his uncles died, bequeathing him the Vatican title of Duke of Camposelice. The vicomtesse was now also a duchess; upstart duchesses may be ignored by nice people, but upstart duchesses with fortunes are usually another matter. They traveled the capitals of Europe in immense and fabulous style and, when they returned to Paris, set up house at No. 27 Avenue Kléber, where the Duke indulged his passion for collecting old musical instruments. "The Duke had nothing but his title when he married Mrs. Singer; but she was wealthy, and they attempted virtually to buy their way into polite society in Paris," remarked one commentator, continuing that "in this they were hardly successful . . ." Isabella might justifiably have retorted that it all depends on what you mean by success.

In many ways, it is true, this marriage was not a success. By marrying, Isabella lost her life interest in the Oldway estate; but this did not leave her a poor woman, since she still held her share of Singer stock, worth about $1.5 million. What was more, all her children except Mortimer, the eldest, who remained in England and was later sent off on a round-the-world voyage accompanied by a tutor in an effort to preserve him from temptation and curb his extravagant habits, went with her to Paris, and they all had tidy amounts of property and incomes of their own.

The Duke made it clear that he considered this to be family income and felt that he was entitled to the use of it. This was naturally a matter of grave concern to the boys' financial advisers and the estate's executors. "I warned Madam and begged her not to marry that man," wailed one of these, George Woodruff (in charge of the

Singer operation in England), in a letter to Hawley after one particularly severe crisis. They were all afraid that the Duke would considerably diminish, if not altogether waste, the money belonging to the children before they came of age. To avoid this, Singer's Paignton solicitor, Yard Eastley, hit upon a solution: as each of the boys reached the age of sixteen he was advised to run away from home, whereupon he would be made a ward of court and his portion would be safe from the grasping hands of his stepfather. It is not clear quite what part Isabella played in these dramas. Although she assured Woodruff that her husband "has always been kind to her and the children," Paris, when he in turn left home, had a different tale to tell. "Paris says they go for weeks without speaking to each other, and he has heard the Duke tell his Mother that he got all the property in his own name and if she did not mind what she done [*sic*] he would turn her out," wrote Woodruff to Hawley. The occasion of Paris's leaving was apparently a stupendous quarrel in which the Duke hit Isabella, Paris intervened and knocked his stepfather down, and pistols were brought out before everyone at length calmed down.

However, this marriage, though so unsatisfactory in some ways, did enable Isabella to achieve some degree of acceptance. She and the Duke gave enjoyable and well-attended musicales at the Avenue Kléber house. The popularity of these gatherings does not seem to have been altogether connected with that of the host since "when he was buried [in 1887] few of those who were so prompt to respond to invitations to the dead man's fetes and concerts thought it worthwhile to attend his funeral." But society is, on the whole, concerned with the living rather than the dead; and by the time the Duke died, Isabella could reasonably feel that as far as her daughters were concerned she was well on the way to achieving her ends. If *le tout Paris* declined to attend the Duke's funeral, it had come to his parties, where the young and pretty Misses Singer were very much in evidence.

The Duke died in September 1887; in July of that year Isabella's eldest daughter, Winnaretta, then aged twenty-two, married the Prince Louis de Scey-Montbéliard. Despite his resounding title the Prince was not much of a catch, although some improvement on Alice's commercial gentleman. "Her marriage to Prince Louis," commented the columnist quoted above, "is supposed to have been the result of a desire to possess a home of her own and protection from the importunities of her mother's impecunious but enterprising husband. Now death has relieved her of the old Duke's presence, she may have occasion to regret her haste." Indeed, by 1889 Isabella was writing to Hawley: "You will be surprised to hear that Winnie is suing for a separation from her husband—it seems they have bothered her so much for money that she won't stand it any longer and he married her with false papers he is no more a Prince than I am so she is paying for her bad behavior to me and getting punished for it—however I do not think she cares one bit about it." By that time, however, Isabella could congratulate herself on what to her was her first real coup: in 1888 Winnie's younger sister Isabelle (known in the family as Belle-Blanche) got engaged, as Isabella proudly wrote Hawley, to "the young Duke Decazes and . . . he belongs to one of the first families in France." In fact Isabella does not seem to have distinguished between the *ancien régime* and the *nouveaux arrivés* to which Elie Decazes, although the son of a celebrated government minister, certainly belonged. But this distinction did not escape other members of society, one of whom, when asked if this was a real duke, replied, "Yes, he is quite real, but 'machine made.'"

To the American press, however, the Duke seemed to be quite a catch. The news made the headlines in New York, and most of the attention was concentrated almost exclusively on the cash aspects of the affair. "What the original cost of the Duke was does not appear," wrote the *World*, "but it is evident already that French noblemen properly married, decorated, appointed, housed and fed are very costly commodities for American heiresses to deal in." Or, as the *Tribune* summed it up in a headline: "SHE PAYS ALL THE BILLS—HE THINKS HIMSELF CHEAP AT THE PRICE."

This was the epoch of marriages between new American capital and ancient but bankrupt European nobility—"ANOTHER AMERICAN DUCHESS," commented the *World* on Blanche's wedding. The New York press, reflecting the social attitudes of Old New York, made much of the lack of respectability of both Blanche's parents. "The future Duchesse Decazes is the youngest daughter of the late Isaac

Singer's third acknowledged wife, who was recognized on Mr. Singer's death to be his legal widow ... In Paris she is described as the daughter of an old-time Parisian boarding-house keeper named Mme. Boyer," wrote the *World*. The French papers, however, had no interest in criticizing their future duchess. As one wrote tactfully, "Je ne prétends point ici faire l'historique des deux familles."

When it came to the actual wedding, though, the social splendors silenced even the most determined carpers. It was everything Isabella could have hoped for—indeed, almost more than in her wildest dreams she could possibly have expected. It was hailed as the social event of the season. More than three thousand people tried (many failed) to get into the church, repairing afterward to 27 Avenue Kléber, where once again Isabella could congratulate herself on having shed a possibly unpresentable husband at the crucial time, retaining only his advantages—in this case her title.

The coverage of the wedding by the American press demonstrates clearly its strange, for a nation of commoners, fascination with titles and privilege.

> Precisely at noon [reported the *Tribune*] the bridal party entered, while the organ pealed forth the joyous strains of Mendelssohn's Wedding March. Queen Isabella and Princess Philip of Saxe-Coburg-Gotha sat on opposite sides of the altar, which was fragrant with flowers and incense. The bride, a beautiful girl of nineteen, with an income of $120,000 a year, wore a superb costume of white peau-de-soie, said to have cost 30,000f. It was trimmed with orange blossom and a tulle veil covered her face. The Duchesse de Camposelice wore a pearl gray brocade silk, with tassels of the same color, and a bonnet of lace and feathers glittering with diamonds. The Duchesse Decazes was radiant in lilac silk, lace and diamonds. The Princesse de Scey-Montbéliard wore pale rose. Princess Philip of Saxe-Coburg-Gotha appeared in pale blue silk, with demi train, trimmed with lace, and a bonnet of lace, pearls, feathers and old rose ribbon. Queen Isabella wore a striped old rose silk, pearls and a Persian mantle.

In such company everyone possible was invested with nobility: in the *World*, Hawley, one of the witnesses, became "d'Hawley" for the occasion.

Sadly this desirable marriage was cut short by Blanche's death in 1896, when she was only twenty-seven. Whether it could ever have taken place at all had her father still been alive is another question. An heiress, whatever her expectations, is no substitute for a woman already wealthy in her own right; and would Singer ever have consented to leave his beloved Wigwam? Moreover, it was easier to gloss over the Duchesse de Camposelice's family anomalies with her first—and indeed her second—husband out of the way. Thus, ironically, the death of the protagonist in good time seems to have been a prerequisite for the final fulfillment, in this instance, of the immigrant's dream. And while Isaac would no doubt have been gratified by the whole affair, especially by the superb wedding and by the idea of his daughter becoming a duchess, it is doubtful whether he would have derived quite the same satisfaction as Isabella from the mere concept of being allied to one of "the first families of France." The mother's and daughter's aspirations, in this case, had passed beyond his.

Singer might have taken more active pleasure in the doings of his other daughter, Winnaretta. The headstrong Winnie, having divorced her unsatisfactory prince, was able to launch herself into French society in her own right as a patron of the arts, and as no mean painter and musician herself. There can be no doubt that French society, with its tradition of intellectual, artistic and musical *salons*, suited her far better than British society, so resolutely anti-intellectual and so dull. As it was, she eagerly took her place in that part of the Faubourg St.-Germain which interested itself in the arts—the set whose activities were so avidly recorded by Marcel Proust, whose friend she became. In 1893 she was married again, this time to Prince Edmond de Polignac, bearer of one of the greatest names in France, a man of intelligence and sensibility, a talented minor composer, a collector of paintings, and a literary figure. The prince being considerably older than she was, this marriage, too, was cut short by his death in 1903; but unlike her first attempt it was very happy while it lasted. The union gave rise to a number of droll stories. Jacques Emile Blanche remembered "the prince jumping over a chair at the Blanches' Dieppe villa, by way of proving he was still young enough to marry, and old Mme. Blanche saying: 'So the lute is going to marry the sewing machine.'" Proust recorded that, before her marriage, Winnaretta "already frequented a most artistic and elegant set . . . her annual salons of paintings always contained some remarkable exhibits and acquisitions."* Of the marriage, he said:

> They were both musicians, and both very drawn to the intellectual life whatever form it took. The only thing was that she was always too hot, while he felt the cold very badly. He was therefore always extremely uncomfortable amid all the drafts in the big studio at the rue Cortambert [the magnificent house built by Winnaretta at Passy]. He protected himself as best he could, always wrapping himself in thick plaids and travelling blankets. When people teased him about this, he would reply, "Well, as Anaxagoras said, life is a journey!"

One aspect of this marriage which particularly pleased the prince was the fact that Winnaretta possessed Monet's magnificent picture "A tulip-field at Haarlem." Before they married, the prince had seen it at a sale and had coveted it; "'But you can imagine,' he would say, 'it was bought by some American woman—I couldn't think of anything bad enough to say about her. A few years later I married her, and now I own the picture after all!'" After the death of her husband, Winnaretta continued as a famous patron of the arts; her salon was particularly notable for the concerts of modern music, some of it commissioned by the Princess, which were performed there. She

---

* However, that curious figure of the Parisian *fin de siècle*, Robert de Montesquiou, was much less gallant about Winnaretta's artistic inclinations. He was very much put out by the Polignac marriage, and in 1893 he wrote an unpleasant satirical poem called "Vinaigrette" in which he accused her of signing her name to what was effectively the work of other people, and then taking the credit:

> . . . un artiste la sert
> Et lui dit où se doit placer le bleu, le vert . . .
> Elle signe, elle envoi au Salon, le medaille
> Est pour elle . . .

However, even Montesquiou, possibly the most malicious man in Paris, did not dare publish this piece.

became a famous Parisian institution. Dying childless, her will established the Fondation Singer-Polignac, which supports science and the arts. Once again, one feels that her father, could he have foreseen this future for his money, would have enjoyed the concept more than the actuality. Winnaretta would have been altogether too intellectual for him, although he would have approved of her artistic endeavors and appreciated her habit of putting on fancy dress whenever the whim took her.

Singer's daughters had thus done very well; but of all the children of this last family, there can be no question that the one who was most truly a man after his father's heart was the third son, Paris.

Of Isabella's four sons, two, Mortimer and Washington, decided to make their principal homes in England, leading lives of unimpeachable sporting respectability after, in the case of Mortimer, a dissolute youth in which everyone despaired of him and confidently expected him to run through his fortune before he was thirty. Despite this unpromising start, he became Sir Mortimer and in 1921 was High Sheriff of Berkshire. His brother Washington was Sheriff of Wiltshire in 1924. (The latter was possibly the only person to hold such a post able to record his birthplace, in *Who's Who*, as "Yonkers, N.Y.") Washington was particularly devoted to horse racing; his most tangible memorial is now the Washington Singer Stakes run every year at Newbury. Franklin, the youngest brother, also had sporting interests, in his case, yachting.

Paris, however, like Winnaretta, seems to have inherited something of his father's artistic inclinations. He also inherited his other, more earthy inclinations, finally combining his taste for the arts with his taste for pretty women in a long and much-publicized liaison with the dancer Isadora Duncan. Paris and Isadora first met at the funeral of the Prince de Polignac, his brother-in-law and her patron, but Isadora did not recall this when, some years later, in 1909, he presented himself in her dressing room after a performance, bent to kiss her hand, and said: "You do not know me, but I have often applauded your wonderful art." A clairvoyant had recently forecast that Isadora would soon meet a millionaire who would rescue her from her incessant debts and fulfill all her dreams, and Paris Singer

was the man. In her memoirs he appears as "Lohengrin"—when she
wrote them he was still very much alive and she was afraid that, if she
were too explicit, he would stop publication. At the time they met,
Paris was separated from his wife, and he and Isadora fell passionately
in love. Paris looked like his father: he was over six feet three inches
tall, with thick blond hair. Stanislavsky, who met them in Paris in
1909, described him as "a very handsome man, taller than I . . . magnificent, businesslike, and with excellent manners."

Isadora, who possessed that peculiarly American capacity to
interpret social and artistic theories seriously and above all literally,
was always, as a devout Communist, most embarrassed by the Singer
riches and the fact that they were hers to dispose of as she wished.
She enjoyed the life-style they provided but felt guilty for doing so.
Consequently, she took every opportunity to show up Paris's essential bourgeois banality of mind beneath his glittering exterior. When,
after the birth of their son Patrick, he proposed that they should
marry, she was shocked—or said she was:

> "How stupid for an artist to be married," I said, "and as I must
> spend my life making tours round the world, how could you spend
> your life in the stage-box admiring me?"
>
> "You would not have to make tours if we were married," he
> answered.
>
> "Then what should we do?"
>
> "We should spend our time in my house in London, or at my
> place in the country . . ."
>
> L. proposed that we should try the life for three months. "If
> you don't like it, I shall be much astonished."

Isadora agreed to try the life, and Paris then took her to The Wigwam
("my place in the country") in which he had earlier bought out his
brothers' interests and which he had remodeled after Versailles and the
Petit Trianon (he had at one point qualified in architecture). However, she still could not bring herself to marry him. Later in her life,
despite the fact that it was against her principles, she did in fact get
married—to Sergei Essenin; but this was ideologically acceptable on
account of his being a Russian Communist and a poet. He was also
penniless, a definite point in his favor. From Isadora's point of view,

however besotted she might be with Paris, there was no getting around the fact of his wealth. "Can you imagine a worse curse," she once said to another friend. "I didn't have to imagine it. I have witnessed it and I am sorry to say even have lived through it. Just imagine how degrading it must be to be identified, not with yourself as a person, with what you have done or are doing, but with your signature on a check . . . It is certainly more difficult for a rich man to accomplish anything serious in life." Isadora took life and herself nothing if not seriously. The kind of temperamental clash that must always have been taking place between her and Singer was typified by his reaction one evening to her recital of Walt Whitman's "Song of the Open Road," which was her favorite poem. "Singer exclaimed, 'What rot! That man could never have earned his living!' 'Can't you see,' Isadora protested, 'he had the vision of Free America?' 'Vision be damned!' Singer exclaimed." Paradoxically, of course, Paris Singer was the ultimate product of the kind of freedom at that time specific to America: the freedom to try any new idea and, if you were lucky, enjoy unlimited wealth as a result of it.

Apart from his affair with Isadora, Paris Singer is known as the inspiration and money behind the building of the new resort of Palm Beach, Florida (before he lost much of his fortune in the stock market crash of 1929). He conceived this in collaboration with the architect Addison Mizner, who became famous for the rococo palaces he built there for his rich patrons (and was the brother of the famous con man and wit, Wilson Mizner). There are various fanciful versions of the genesis of Palm Beach as we know it and as Singer and Mizner built it.

One morning in January of the year 1917 Paris Singer and Addison Mizner sat rocking at the porch of the old Royal Poinciana [goes one version]. Both were bored. "Mizner," said Singer abruptly, "you know I came here expecting to die, but I'm damned if I feel like it." Mizner shrugged. "What are you going to do about it?" he asked. It was Singer's turn to shrug. "What would you do about it," he said, "if you could do anything you wanted?" Mizner suddenly became earnest. With a sweeping gesture he took in Palm Beach's Poinciana, its Breakers, its station and its half a hundred shingled cottages. "I'll tell you what I'd do," he said. "I'd build something that wasn't made of wood, and I wouldn't paint it yellow."

Another source attributes it all to Isadora, albeit indirectly:

Any disappointment in a romantic matter always caused [Singer] to console himself with architecture. A tiff was enough to start him on a villa or a harbour improvement. A broken heart inspired a project for a great medical research institute, but a reconciliation cancelled it. The heroine of Singer's epic romance was the celebrated dancer Isadora Duncan . . . Their chief difference arose over her enthusiasm for handsome young fellows. She had a weakness for prizefighters . . . The curtain was finally rung down on the world-famous romance when Isadora, during a vacation in Palm Beach, happened to catch a glimpse of the handsome physical instructor of Gus's Baths. Paris, returning to Palm Beach from a business trip, discovered that Kid McCoy had not only ousted him from his place in Isadora's heart but also formed a habit of throwing big champagne parties and charging them to Paris Singer's account. The sewing machine heir told the dancer that he was discarding her for ever. She retorted with an Irish curse guaranteed to make him lose all his money. He called in Addison Mizner and started to build and build and build. He didn't stop until the skyline and landscape of the enchanted isle were completely changed. The new Palm Beach . . . was Paris Singer's substitute for Isadora Duncan.

In many ways Paris's life was a reenactment of his father's, but on a much grander scale. Isaac knew Julia Dean's parents; Paris made one of the most celebrated theatrical figures of the day his mistress. Isaac ran the Merritt Players from a one-horse wagon; Paris bought Madison Square Garden for Isadora (she turned it down). Isaac built The Castle and The Wigwam but could never manage to lure the right people to his entertainments there; Paris built Palm Beach and became the social tyrant of the Everglades Club. In him the capitalist romance—in all senses—found its apotheosis.

But while this social transformation was going on, the figure of Isaac Singer himself was curiously transmogrified. For example, in Washington's obituary in the *Torquay Times* he appears as "the late

Sir Isaac Singer." When Alice, under the name of Agnes Leonard, first appeared on the English stage in 1878, *The Times*, which gave her an excellent write-up ("Occasionally we find a beginner who makes for himself a conspicuous place on the stage and is from the very first regarded as an actor of undoubted ability. One of these lucky few is now before us in the person of Miss Agnes Leonard . . ."), said of her that "Miss Leonard . . . comes of a good old English family" which emigrated to New York (and it gave her date of birth as 1856; Alice presumably felt that twenty-two was a more suitable age for a debut than twenty-seven). But the greatest inaccuracy of all emanated from Paris via Isadora. Toward the end of her life, Isadora's friend Victor Seroff asked her whether it was in fact Paris Singer who invented the sewing machine. In reply, Isadora recounted the one story which she remembered Paris telling her from their years together: "'Of course not, I wish he had. It was not even old Isaac, his father, but his grandfather who invented it,'" she declared to Seroff.

"It seems that he struggled with it for years, because he could not solve the most important problem. Since a person sews with a needle, he first pierces the cloth with it and then, letting go of the needle, catches it again with his hand on the other side of the cloth. But the machine did not have such a hand . . ." I never did know [continued Seroff] whether the romantic solution of the problem which she told me then was a product of Isadora's or of Paris Singer's imagination, but according to Isadora the old man had a dream in which "a rider on a horse was rushing at him with a spear in his hand. The spear had a hole at its point, threaded with yellow ribbon, which waved in the wind. The problem was solved: the hole for the thread had to be changed to the point-end of the needle."

It was a story which the sewing machine's inventor might not have recognized. But he would definitely have appreciated it.

Notes

Chapter 1: Early Influences

p. 6   One of his granddaughters . . .: This is the account of Winnaretta Lady Leeds, Isaac's only surviving grandchild, now resident at Paignton, Devon.

—There is another theory . . .: According to François Roche de Vercors, who has made a genealogical study of various American families.

—*National Cyclopedia of American Biography*: 1943, XXX, p. 544.

p. 7   In September of that year . . .: Account in Orsamus Turner, *History of the Pioneer Settlement of Phelps and Gorham's Purchase*. The Oneidas sold their lands for $2,000 cash, $2,000 in clothing and other goods, $1,000 in provisions, $500 in money for the erection of a saw-mill and grist-mill on their reservation, and an annuity of $100 in silver forever. The Cayugas sold all except a reservation of one hundred square miles for $500 in hand, $1,628 to be paid the following June, and an annuity of $500 forever.

—". . . progressed far beyond the bounds of civilization": Ibid.

p. 8   "the weight of flour . . .": John Fowler, in *Journal of a Tour in the State of New York in the year 1830*, wrote: "Within the limits of the village [of Rochester] are eleven flouring mills, containing 53 run of stones, capable of manufacturing 2,500 bushels of flour, and consuming more than 12,000 bushels of wheat every twenty-four hours. Some of the mills are of a scale of magnitude perhaps not equalled in the world, and all are considered unrivalled in the perfection of their machinery."

—"Mechanics are . . .": John Melish, *Travels in the United States of America . . .*

p. 9   . . . land was naturally much more expensive . . .: In his *Journal of a Year's Residence in America, 1819*, William Cobbett estimated the price of average farming land, already cleared, "the land fenced into fields with posts and rails, the woodland being in the proportion of one to ten of the arable land, and there being on the farm a pretty good orchard," at sixty dollars an acre. In Oswego it might have been more expensive still: "the rich lands on the *necks* and *bays*, where there are meadows and surprisingly productive orchards, and where there is *water carriage*, are worth, in some cases, three times the price," comments Cobbett.

p. 10 ". . . without money . . .": Almost all the detail available about I. M. Singer's childhood comes from a newspaper interview which he gave to the *Atlas* of New York as part of their series "Portraits of the People" on 20 March 1853, when he was just beginning to make a name for himself on account of the sewing machine.

—"It is, indeed, scarcely credible . . .": J. Fowler, op. cit.

p. 11 "schools at that day . . .": *Atlas*, 20 March 1853.

—"In 1800 . . .": Quoted in Turner, *History of . . . Phelps and Gorham's Purchase*.

p. 12 ". . . a common school . . .": *Atlas*, 20 March 1853.

p. 14 "The woman told me . . .": Frances Trollope, *Domestic Manners of the Americans*.

Chapter 2: The Strolling Player

p. 16 "Usually the characters . . .": Quoted by Richardson L. Wright in *Hawkers and Walkers in Early America*.

—Playbill quoted by John Bernard in *Retrospections of America*.

p. 17 . . . historian of the town . . .: John Fowler, *Journal of a Tour in the State of New York in the year 1830*.

—"Sam Patch . . .": Ibid.

p. 18 Dean asked him . . .: These recollections of Dean's are quoted by Tom Mahoney in "One Hundred Million Sewing Machines," commissioned by the Singer Company but never published.

—. . . the most likely available reading material . . .: These phenomena are noted by Louis B. Wright in the chapter on "The Cultural Repertory of the Frontier" in his *Culture on the Moving Frontier*.

p. 20 "Most of his time . . .": Evidence given during the court case over Singer's will in 1875.

p. 21 "His intimacy . . .": Where no source is given for a quotation, it has been taken from one of the many undated and unattributed press clippings relating to Singer, collected in a scrapbook by his executor, David Hawley. The scrapbook is now held by the Singer Company's New York office.

p. 25 ". . . operated . . . by a crank . . .": U.S. Patent no. 1,151, 6 May 1839.

p. 27   Sol Smith . . .: From Sol Smith, *Theatrical Management in the West and South.*

p. 28   In 1832 . . .: This estimate was made by Judge Cranch of Washington City and is quoted by George Barrell Cheever in *The True History of Deacon Giles' Distillery.*

—"The dreadful power . . .": From the chapter on temperance literature in Edmund Pearson, *Queer Books.*

p. 29   "There were two men . . .": Quoted by A. M. Earle in *Stagecoach and Tavern Days.*

—". . . more than double that number . . .": Bernard, *Retrospections of America.*

p. 30   "The cause . . .": Trollope, *Domestic Manners of the Americans.*

—Sol Smith recalled . . .: Smith, op. cit.

—"Had supper this evening . . .": Quoted by Carl Bode in *Anatomy of American Popular Culture.*

p. 33   "a large quantity of type . . .": *Atlas*, 20 March 1853.

Chapter 3: The Sewing Machine

p. 35   "Everything new . . .": Quoted by H. J. Habakkuk in *American and British Technology in the Nineteenth Century.*

p. 36   "The greatest difficulty . . .": Frances Trollope, *Domestic Manners of the Americans.*

p. 37   "All grain . . .": Habakkuk, op. cit.

—"Restlessness of character . . .": Ibid.

p. 41   "They do far more . . .": Letter in "Ryland's Iron Trade Circular," in Charlotte Erickson, *American Industry and the European Immigrant, 1860–1885.*

p. 43   Mrs. Phelps . . .: Testimony given in the patent suit Singer & Co. *vs.* Walmsley, Baltimore, Md., January 1860.

—"He said," asserted Phelps . . .: Ibid.

p. 45   "Singer talked with me . . .": This, like all other quotations attributed to Zieber, comes from an unpublished memoir of his association with Singer, written after he had finally severed all connections with I. M. Singer & Co. The gist of this memoir, which should always be borne in mind, was that Zieber had been unjustly treated and cheated by Singer and Clark. This was, to some extent, undoubtedly true.

p. 47   "I worked day and night . . .": James Parton, *The History of the Sewing Machine.*

p. 49   Jott Grant . . .: Testimony in Singer *vs.* Walmsley.

p. 50   . . . an advertisement . . .: *Boston Daily Times*, 8 November 1850.

p. 51   Blodgett . . .: Quoted by Zieber.

Chapter 4: Many Inventors

p. 53 On the contrary . . .: The best account of the invention and development of the sewing machine is given in Grace Rogers Cooper, *History of the Sewing Machine.*

p. 54 "Every workman . . .": Parliamentary Papers, 1854–55, XXXVI, p. 146, quoted by H. J. Habakkuk in *American and British Technology in the Nineteenth Century.*

p. 55 "An Entire New Method . . .": Cooper, *History of the Sewing Machine.*
—"Examining some patents . . .": *Journal of Domestic Appliances and Sewing Machine Gazette*, quoted by Brian Jewell, *Veteran Sewing Machines.*

p. 56 Thimonnier's first machine . . .: On view at the Science Museum, South Kensington, London.

p. 57 An argument now arose . . .: Centenary article on Thimonnier in *Bulletin de la Société d'Encouragement de l'Industrie Nationale*, vol. 130, no. 2 (1931).
—"A little further on . . .": Quoted by Jewell, op. cit.

p. 59 "I made that sewing machine . . .": Quoted by F. Lewton in *The Servant in the House.*
—. . ."acquired an aversion . . .": Biography given in O. L. Dick's edition of John Aubrey's *Brief Lives.*

p. 60 "The shop . . .": James Parton, *The History of the Sewing Machine.*
—"It ought . . .": Aubrey, op. cit.

p. 61 ". . . to lie in bed . . .": Parton, op. cit.
—". . . at the present time . . .": Ibid.

p. 62 "I believe . . .": Ibid.
—"His conception . . .": Ibid.

p. 63 ". . . much faster . . .": Ibid.

p. 65 . . . catering for themselves . . .: The appalling conditions under which steerage passengers traveled across the Atlantic at this time are described by Terry Coleman in *Passage to America.* On the whole, the conditions on the America–Europe run were better than in the opposite direction, since the ships were less crowded.

p. 66 "Before his wife left London . . .": Parton, op. cit.
—". . . extremely downcast and worn": Ibid.

Chapter 5: The Sewing Machine War

p. 67 . . . figures prepared by the Wheeler and Wilson Company . . .: *New York Daily Tribune*, 23 May 1862.
—"We know of no class . . .": *New York Herald*, 11 June 1853.

p. 69 "In the last week of June . . .": Karl Marx, "The Working Day," *Das Kapital*, F. Engels, ed.

p. 70 "They might well imagine . . .": Reported in *I. M. Singer & Co.'s Gazette*, 1 November 1859.

—"I met with continual objections . . .": From "Statement of receipts and expenditures by I. M. Singer on account of his invention claimed in his patent re-issued the 3rd day of October 1854." Extract in possession of Lady Leeds.

p. 71 . . . a poster . . .: James Parton, *The History of the Sewing Machine*.

p. 72 "he litigated his way . . .": "Who Invented the Sewing Machine?" in *The Galaxy*, 1867.

p. 73 Where Phineas T. Barnum . . .: After spending some years exhibiting freaks of various sorts around the United States, Barnum presented the Swedish singer Jenny Lind, the "Swedish Nightingale," in two nationwide tours in 1850 and 1851. Lind was a *succès fou*, earned huge sums of money, and made Barnum's reputation as a showman.

p. 74 ". . . at the time . . .": Testimony in the patent suit Singer & Co. *vs.* Walmsley, Baltimore, Md., January 1860.

p. 75 ". . . took occasion to speak . . .": From "Statement of receipts . . .", op. cit.

p. 76 "It is not forty years . . .": *Houston Chronicle*, 4 November 1923.

p. 78 "I asked him why . . .": Testimony in Singer *vs.* Walmsley.

—"Phelps, who was an intemperate . . .": Quoted by Zieber.

p. 80 "The manufacture . . .": From "Statement of receipts . . .", op. cit.

—"Our I. M. Singer . . .": Letters originally in Singer letterbooks, quoted by Tom Mahoney in "One Hundred Million Sewing Machines."

p. 81 ". . . it was absolutely necessary . . .": From "Statement of receipts . . .", op. cit.

p. 87 Thomas Jones . . .: Testimony in Singer *vs.* Walmsley.

p. 89 . . . by 1853 the four biggest . . .: Sewing machine production in 1853, according to Brian Jewell, *Veteran Sewing Machines*, was about 2,550 machines: American Magnetic Sewing Machine Company (est. 1853), 40; Abraham Bartholf (making Lerow and Blodgett machines under license from Elias Howe), 135; Grover and Baker Sewing Machine Company (under license from Howe), 637; Nehemiah Hunt (making machines on Christopher Hodgkin's 1852 patent), about 100; Nichols and Bliss (under license from Howe on Howe's patent), 28; I. M. Singer & Co., 810; and Wheeler and Wilson Company (under license from Howe), 799.

p. 91 "From the outset . . .": John Scott, *Genius Rewarded*.

p. 93 "William Whiting . . .": Argument of George Gifford for Elias Howe in patent renewal suit, 1860.

p. 94 "We take a positive position . . .": *Scientific American*, 1 October 1853.

p. 95　"... had a dislike ...": *The Galaxy*, 1867.

—"...just when it was about to yield ...": Parton, op. cit.

p. 96　"Perhaps thirty ...": Ibid.

—"He had suits pending ...": Testimony, Singer *vs*. Walmsley.

p. 99　"It is not within the history of invention ...": *New York Daily Tribune*, 10 August 1860.

Chapter 6: The Beginning of Mass Production

p. 100　"Singer & Co. were struggling ...": John Scott, *Genius Rewarded*.

p. 101　The great transition ...: Production figures quoted in the *New York Times*, 7 January 1860.

—Singer machines in 1850 ...: According to Singer, quoted in the *New York Daily Graphic*.

—By 1870 ...: Figures given by Frederick Bourne in "American Sewing Machines" in *One Hundred Years of American Commerce*, vol. 2, ed. Chauncey M. Depew.

p. 102　The armaments industry ...: For a detailed description of these developments, see L. T. C. Rolt, *Tools for the Job*, and J. W. Roe, *English and American Tool Builders*.

p. 103　"I put several ...": *Writings of Thomas Jefferson*, ed. H. A. Washington, vol. 1, quoted in Rolt.

—"In the same year ...": Rolt, op. cit.

p. 104　"The several parts ...": Ibid.

p. 106　... the first milling machine ...: By Frederick Webster Howe for the Robbins and Lawrence Company of Windsor, Vermont.

p. 107　"On four lots of ground ...": *I. M. Singer & Co.'s Gazette*, December 1857.

p. 108　"Let us suppose ...": *New York Daily Tribune*, 23 May 1862.

p. 109　"... in the United States ...": *Mechanic's Journal*, 1 April 1858.

p. 110　"... the workmen ...": Parliamentary Papers 1854, XXXVI, p. 146, quoted by H. J. Habakkuk in *American and British Technology in the Nineteenth Century*.

Chapter 7: Opening the Market

p. 111　"Isaac M. Singer ...": *Atlas*, 20 March 1853.

—"This is a country ...": Calvin Colton, *Junius Tracts*.

—Harper & Brothers ...: Charles Seymour, *Self-Made Men*.

p. 112　"One thinks sometimes ...": George Templeton Strong, *Diaries 1835–75*.

—"Mr. I. M. Singer ...": *Scientific American*, 11 August 1855.

p. 114　"Two hundred and sixty ...": Captain Marryat, *A Diary in America, with Remarks on its Institutions*.

p. 115 Edward Clark could write . . .: Letters quoted by Andrew B. Jack, "The Channels of Distribution for an Innovation: The Sewing Machine Industry in America 1860–65," in *Explorations in Entrepreneurial History*.

p. 120 "Mr. Plumley . . .": *Frank Leslie's Illustrated Weekly*, 30 July 1859.

p. 121 ". . . the women do not seem . . .": Roger Burlingame, *Machines That Built America*.

—"A most Wonderful Invention . . .": *Harper's Weekly*, 10 March 1866.

p. 122 Newspapers printed letters . . .: *San Francisco Daily Evening Bulletin*, 21 November 1860.

p. 127 The point has been made . . .: For instance, by Burlingame in *Machines That Built America*.

p. 132 ". . . soft sawder and human natur' . . .": Richardson L. Wright, *Hawkers and Walkers in Early America*.

p. 133 The insistence of these agents . . .: The "Rules and Regulations of the Buffalo Sewing Machine Union relating to Canvassers" may be of some interest in this context:

1. The time for payment of machines valued at sixty-five ($65) dollars or less, shall be limited to one year; all over that price fifteen months. The first payment to be ten dollars wherever practicable, and in no case less than five dollars; and all notes or contracts to be made with interest from date, a discount of seven percent only to be allowed to cash purchasers.

2. No Agent, Canvasser or employee of any company hereby represented shall allow a machine to be taken to any locality to interfere or compete with any other machine that may be placed there for trial, provided that said trial does not *exceed* ten days, that being the full term allowed for that purpose. During that time no interference with or misrepresentation of such machine shall be allowed under any circumstance whatever.

During the time of said trial the machine may be exchanged at the option of the company or agent; but at the end of said time, if the machine is not sold, it shall be removed.

There shall be no interference with any machine that is sold and not paid for.

3. No machine or extra attachments shall be sold for less than the printed list price of the company so selling; and no extra attachments will be given with machines without additional payment for the same at regular prices, as follows: Tucker, $3.00; Corder, $2.00; Common Ruffler, $1.00 (when sold with machine, the three may be sold for $5.00); Johnston Ruffler, $2.00; Harris Binder, $1.50; Bed Hemmer, $1.00.

4. No payment of five dollars, more or less, shall be made to outside

parties, for customers of machines, unless said parties will become responsible for, and give the entire instruction on the machine.

5. Any member of this Association, having in his employ anyone who shall be found guilty of violating any of the rules and regulations of the Association, shall pay into the treasury thereof the sum of ten dollars for the first offence, for the second offence committed by the same employee, twenty-five dollars, and for the third offence, said employee shall be discharged from his situation.

Signatories for: Weed Sewing Machine, Victor Sewing Machine, Singer Sewing Machine, Aetna Sewing Machine, Original Howe Sewing· Machine, Grover and Baker Sewing Machine, Improved Elias Howe Sewing Machine.

p. 135 . . . sales continued to rise . . .: Figures given in *Practical Mechanic's Journal* from a series of articles on the sewing machine, 1860–62.

—"So lately is it . . .": *Practical Mechanic's Journal*, 1 April 1858.

Chapter 8: Mr. Mathews, Mr. Merritt and Mr. Singer

p. 149 "John Tomkins . . .": Quoted by Bernard, *Retrospections of America*.

—"Ours is a country . . .": Calvin Colton, *Junius Tracts*.

—"How New York has fallen off . . .": George Templeton Strong, *Diaries 1835–75*.

p. 150 "There was in all the tribe . . .": Edith Wharton, "The Old Maid" in *Old New York*.

—"These fashionable parties . . .": Washington Irving, *A History of New York*.

—. . . Ward McAllister . . .: 1827–95. The arbiter of New York society during the second half of the nineteenth century. Having made a fortune as a lawyer and married an heiress, he devoted himself entirely to social life, which, he considered, embodied all those graces and refinements which make civilized life worth living. He is still remembered for the classifications he imposed upon society. In 1872 he brought into being the "Patriarchs" as a protest against the powers of exclusion held by a few very rich men. To counter this exclusion, he banded together the oldest New York families, whose approval of any social aspirant was an absolute necessity; the heads of these families were dubbed the Patriarchs. He is perhaps best known for the creation of the "Four Hundred." In 1892, when Mrs. William B. Astor found that her ballroom would not accommodate all the people on her list, McAllister offered to cut the list down for her, reducing it to the four hundred who would fit into the room; he afterwards boasted at the Union Club that there were "only about four hundred people in New York society."

p. 151 "The delicate descendant . . .": "The Editor's Easy Chair," *Harper's Magazine*, September 1860.

p. 152 ". . . one worried about being 'showy' . . .": Stephen Birmingham, *Our Crowd*.

—". . . the style of everyday living . . .": T. C. Grattan, *Civilised America*.

p. 154 They would put their cash . . .: See E. J. Hobsbawm, *The Age of Capital 1848–75*.

—This is not to say . . .: For the full story of this romance, see Johanna Johnston, *Mrs. Satan*, the biography of Tennessee's sister, Victoria Woodhull.

p. 156 "The most remarkable equipage . . .": *New York Herald*, 5 December 1859.

p. 157 . . . a dashing five-in-hand . . .: *New York Family Herald*, 22 June 1859.

p. 158 . . . Jay Gould . . .: 1831–92. Perhaps the most notorious of the "robber barons" who amassed huge fortunes by dubious means. In Gould's case these included flooding the market with worthless shares, watering stock, bribing legislators to get laws passed which were favorable to himself. He first came to the public eye with the fight, which he conducted in association with Jim Fisk and Daniel Drew, against Cornelius Vanderbilt to gain control of the Erie Railroad. Before this he had already driven at least one business partner to suicide. Gould's and Fisk's attempt to corner the gold market brought about the Black Friday panic of 24 September 1869; this led to a good deal of public anger directed against Gould. Unabashed he went on to make another fortune in western railroads, and died of tuberculosis in 1892. He was a solitary man, his only non-financial interests being books and gardening.

Chapter 9: The Castle and The Wigwam

p. 180 "Maybe I was . . .": *Wall Street Journal*, 30 March 1927.

p. 187 "Our dear Father . . .": Letter to Winnaretta, Princesse de Polignac, 4 January 1906.

—". . . as far as possible . . .": This and many of the subsequent descriptions of The Wigwam and life in it are taken from some unpublished "Reminiscences of Oldway and Paignton" by George Bridgman's daughter, Mrs. Laura Goss, who knew the family well.

—". . . it does not possess any marked architectural features . . .": C. H. Patterson, *History of Paignton*.

p. 189 "The parties and balls . . .": Ibid.

Chapter 10: "A Very Ghastly Domestic Story"

p. 199 "I expect William . . .": All details of the will settlements, Isabella's letters and many of the newspaper reports were preserved by David Hawley and are now in the possession of the Singer Company, New York.

Chapter 11: The Immigrant's Dream Fulfilled?

p. 212 "The Duke had nothing . . .": All undated quotations from newspapers are taken from clippings found in David Hawley's collection.

p. 214 But this distinction . . .: According to Winnaretta, Lady Leeds.
—"SHE PAYS . . .": *New York Herald Tribune*, 29 March 1888.

p. 216 ". . . the prince jumping . . .": George D. Painter, *Proust*, vol. 1.

p. 217 Proust recorded . . .: Marcel Proust ("Horatio"), "Le Salon de Princesse Edmond de Polignac," *Le Figaró*, 6 September 1903.
—"But you can imagine . . .": Proust, op. cit.
—". . . un artiste la sert . . .": Quoted by Michel de Cossart in "Princesse Edmond de Polignac: patron and artist," *Apollo*, August 1975.

p. 218 "You do not know me . . .": Isadora Duncan gave an account of this episode in *My Life*; Victor Seroff gives a possibly more reliable one in *The Real Isadora*.

p. 219 "How stupid . . .": Duncan, *My Life*.

p. 220 "Can you imagine . . .": Seroff, *The Real Isadora*.
"Singer exclaimed . . .": Ibid.
—"One morning in January . . .": Cleveland Amory, "The Old Palm Beach," *Diners Club Magazine*, July 1961.

p. 221 "Any disappointment . . .": Alva Johnston, *The Incredible Mizners*.

p. 222 "Of course not . . .": Seroff, *The Real Isadora*.

Bibliography

Arthur, Timothy Shay. *Ten Nights in a Bar-room*. New York: 1854.

Aubrey, John. *Brief Lives*, ed. O. Lawson Dick. London: Secker & Warburg, 1969.

Auchincloss, Louis. *The Embezzler*. London: Victor Gollancz, 1966.

Bernard, John. *Retrospections of America*. New York: Harper & Brothers, 1887.

Birkbeck, Morris. *Notes on a Journey in America*. London, 1819.

Birmingham, Stephen. *Our Crowd*. New York: Harper & Row, 1967. London: Longman Group, 1968.

Bode, Carl. *Anatomy of American Popular Culture*. Berkeley, Cal.: University of California Press, 1959.

Branca, Patricia. *Silent Sisterhood: Middle-Class Women in the Victorian Home*. London: Croom Helm, 1975.

Brodie, Fawn M. *No Man Knows My History, a Biography of Joseph Smith*. London: Eyre & Spottiswoode, 1963.

Bulletin de la Société de'Encouragement de l'Industrie Nationale. Vol. 130, no. 2 (1931).

Burlingame, Roger. *The American Conscience*. New York: Alfred A. Knopf, 1957.

———. *Machines That Built America*. New York: Harcourt, Brace & Co., 1953.

———. *March of the Iron Men*. New York: Charles Scribner's Sons, 1949.

Cheever, George Barrell. *The True History of Deacon Giles' Distillery*. New York, 1844.

Chevalier, Michael. *Society, Manners and Politics of the United States*. Boston, 1839.

Clark, D. K. *The Exhibited Machinery of 1862—Cyclopedia of machinery represented at the International Exhibition*. London, 1862.

Cobbett, William. *A Journal of a Year's Residence in America, 1819*, London, 1820.

Cochran, Thomas C., and William Miller. *The Age of Enterprise: A Social History of Industrial America.* New York: The Macmillan Company, 1943.

Coleman, Terry. *Passage to America.* London: Hutchinson & Company, 1973.

Colton, Calvin. *Junius Tracts.* New York, 1844.

Cooper, Grace Rogers. *History of the Sewing Machine.* Washington, D.C.: Smithsonian Bulletin no. 254, 1968.

Curti, Merle E., and others, eds. *History of American Civilization.* New York: Harper & Brothers, 1953.

Darby, William. *A tour from the city of New York to Detroit.* New York, 1819.

Delderfield, E. R. *The Torbay Story.* Exmouth, Devon, 1951.

Depew, Chauncey M., ed. *One Hundred Years of American Commerce,* vol. 2. New York, 1895.

Dickens, Charles. *American Notes* (1842). London: Penguin Books, 1972.

Duncan, Isadora. *My Life.* New York: Horace Liveright, 1927.

Dutcher, John W. "The Mysterious Can" (temperance poem). Amenia, N.Y., 1854.

Earle, A. M. *Stagecoach and Tavern Days.* New York: The Macmillan Company, 1900.

Erickson, Charlotte. *American Industry and the European Immigrant, 1860–1885.* Cambridge, Mass.: Harvard University Press, 1957.

————. *Invisible Immigrants: Adaptation of English and Scottish Immigrants in Nineteenth-Century America.* London: Weidenfeld & Nicolson/LSE, 1972.

Ewers, W., and N. Baylor. *Sincere's History of the Sewing Machine.* Phoenix, Ariz.: Sincere Press, 1970.

Fearon, C. H. B. *A Narrative of a Journey of Five Thousand Miles through America.* London, 1818.

Fish, Carl Russell. *The Rise of the Common Man (1830–1860).* Vol. VI, *History of American Life,* ed. Arthur M. Schlesinger and Dixon Ryan Fox. New York: The Macmillan Company, 1927.

Fowler, John. *Journal of a Tour in the State of New York in the year 1830.* London, 1831.

Frost, J. A. *Life on the Upper Susquehanna 1783–1860.* New York: Kings Crown Press, 1951.

The Galaxy. Vol. 4 (31 August 1867).

Grattan, T. C. *Civilised America.* London, 1859.

Habakkuk, H. J. *American and British Technology in the Nineteenth Century.* Cambridge: Cambridge University Press, 1962.

Haight, Gordon. *Mrs. Sigourney: The Sweet Singer of Hartford.* New Haven, Conn.: Yale University Press, 1930.

Hall, Basil. *Travels in North America in the Years 1827 and 1828.* Edinburgh, 1829.

Handlin, Oscar. *The American People.* London: Hutchinson & Company, 1963.

Hansen, Marcus L. *The Atlantic Migration 1607–1860*. Cambridge, Mass.: Harvard University Press, 1940.

Hobsbawm, E. J. *The Age of Capital 1848–75*. London: Weidenfeld & Nicolson, 1975.

Hoyt, Edwin P. *The Vanderbilts*. London: Frederick Muller, 1963.

Irving, Washington. *A History of New York from the beginning of the World to the end of the Dutch dynasty, by D. Knickerbocker*. New York, 1809.

Jack, Andrew B. "The Channels of Distribution for an Innovation: The Sewing Machine Industry in America 1860–65." From *Explorations in Entrepreneurial History*. Cambridge, Mass.: Harvard Business School, n.d.

Jewell, Brian. *Veteran Sewing Machines*. Newton Abbot, Devon: David & Charles, 1975.

Johnston, Alva. *The Incredible Mizners*. London: Hart Davis MacGibbon, 1953.

Johnston, Johanna. *Mrs. Satan: The Incredible Saga of Victoria Woodhull*. New York: G. P. Putnam's Sons, 1967. London: Macmillan, 1967.

Kidder, Jerome. *The Drama of Earth* (temperance). New York, 1857.

Kobler, John. "Mr. Singer's Money Machine," *Saturday Evening Post*, 1951.

Lewton, F. *The Servant in the House*. Washington, D.C., 1929.

Lundberg, Ferdinand. *America's Sixty Families*. New York: Vanguard Press, 1937.

Lyon, Peter. "Isaac Singer and His Wonderful Machine," *American Heritage*, 1963.

Mahony, T. "The Machine That Sews Sews Everywhere," *Readers Digest*, January 1951.

Mahony, T., and Sloane, L. *The Great Merchants*. New York: Harper & Row, 1974.

Manheim, Frank. "The Singer Saga," *Town and Country*, December 1941.

Marryat, Captain. *A diary in America, with Remarks on its Institutions*. Philadelphia, 1839.

Martineau, Harriet, *Society in America*. London, 1839.

Marx, Karl. *Das Kapital*. F. Engels, ed., 1887. As *Capital*. New York: International Publishers Company, 1967.

Melish, John. *Travels in the United States of America in the years 1806, 1807, 1809, 1810 and 1811*. Philadelphia, 1812.

Nevins, Allen (ed.). *American Social History as Recorded by British Travellers*. London: George Allen & Unwin, 1924.

Oswego Palladium, 1819–23.

Painter, George. *Marcel Proust, a Biography*. 2 vols. London: Chatto & Windus 1961 and 1965.

Parton, James. *The History of the Sewing Machine*. Lancaster, Pa., 1868.

Patterson, C. H. *History of Paignton*. n.d.

Pearson, Edmund. *Queer Books*. New York: Garden City Press, 1928.

Practical Mechanic's Journal. London, 1858–63.

Roe, J. W. *English and American Tool Builders*. New Haven, Conn.: Yale University Press, 1916.

Rolt, Lionel T. C. *Tools for the Job*. London: B. T. Batsford, 1965.

———. *Victorian Engineering*. London: Allen Lane, 1970.

Scott, John. *Genius Rewarded* (booklet). New York: Singer Manufacturing Co., 1878.

Seroff, Victor. *The Real Isadora*. London: Hutchinson & Company, 1972.

Seymour, Charles. *Self-Made Men*. New York: Harper & Brothers, 1858.

Singer, C., E. J. Holmyard, A. R. Hall, and T. J. Williams, eds. *A History of Technology, vol. 4. 1750–1850*. London and New York: Oxford University Press, 1958.

Smiles, Samuel. *Self-help* (1859). Centenary ed. London: John Murray, 1959.

Smith, Sol. *Theatrical Management in the West and South*. New York: Harper & Brothers, 1868.

Strong, George Templeton. *Diaries 1835–75*, eds. Allen Nevins and Milton Halsey Thomas. New York: The Macmillan Company, 1952.

Thomson, Gladys Scott. *A Pioneer Family—the Birkbecks in Illinois 1818–27*. London: Jonathan Cape, 1953.

Trollope, Frances. *The Domestic Manners of the Americans*. London, 1832.

Turner, Orsamus. *History of the Pioneer Settlement of Phelps and Gorham's Purchase*. Rochester, N.Y., 1851.

Wallace, Irving. *The Fabulous Showman, a Biography of P. T. Barnum*. London: Hutchinson & Company, 1960.

Weems, Mason Locke (Parson). *The Drunkard's Looking Glass*. n.d.

Wharton, Edith. *The Age of Innocence*. New York: D. Appleton & Company, 1920. London: Constable & Company, 1966.

———. "The Old Maid" in *Old New York*. New York: D. Appleton & Company, 1924.

Wilson, Edmund. *Upstate*. London: Macmillan, 1972.

Woodbury, R. S. "The Legend of Eli Whitney." *Technology & Culture*, vol. 1, no. 3 (1960).

Wright, Louis B. *Culture on the Moving Frontier*. Bloomington, Ind.: Indiana University Press, 1955.

Wright, Richardson L. *Hawkers and Walkers in Early America*. Philadelphia: J. B. Lippincott Company, 1927.

Index

2 - 5 pts.
EQUIPAGE

50 - Strong